ALISON POULIOT is an ecologist and environmental photographer with a focus on fungi. Her journeys in search of fungi span northern and southern hemispheres, ensuring two autumns and a double dose of fungi each year. Alison is the author of *Underground Lovers* (NewSouth, 2023, published as *Meetings with Remarkable Mushrooms* by Chicago University Press, 2023), *The Allure of Fungi* and co-author of *Wild Mushrooming*.

Funga Obscura

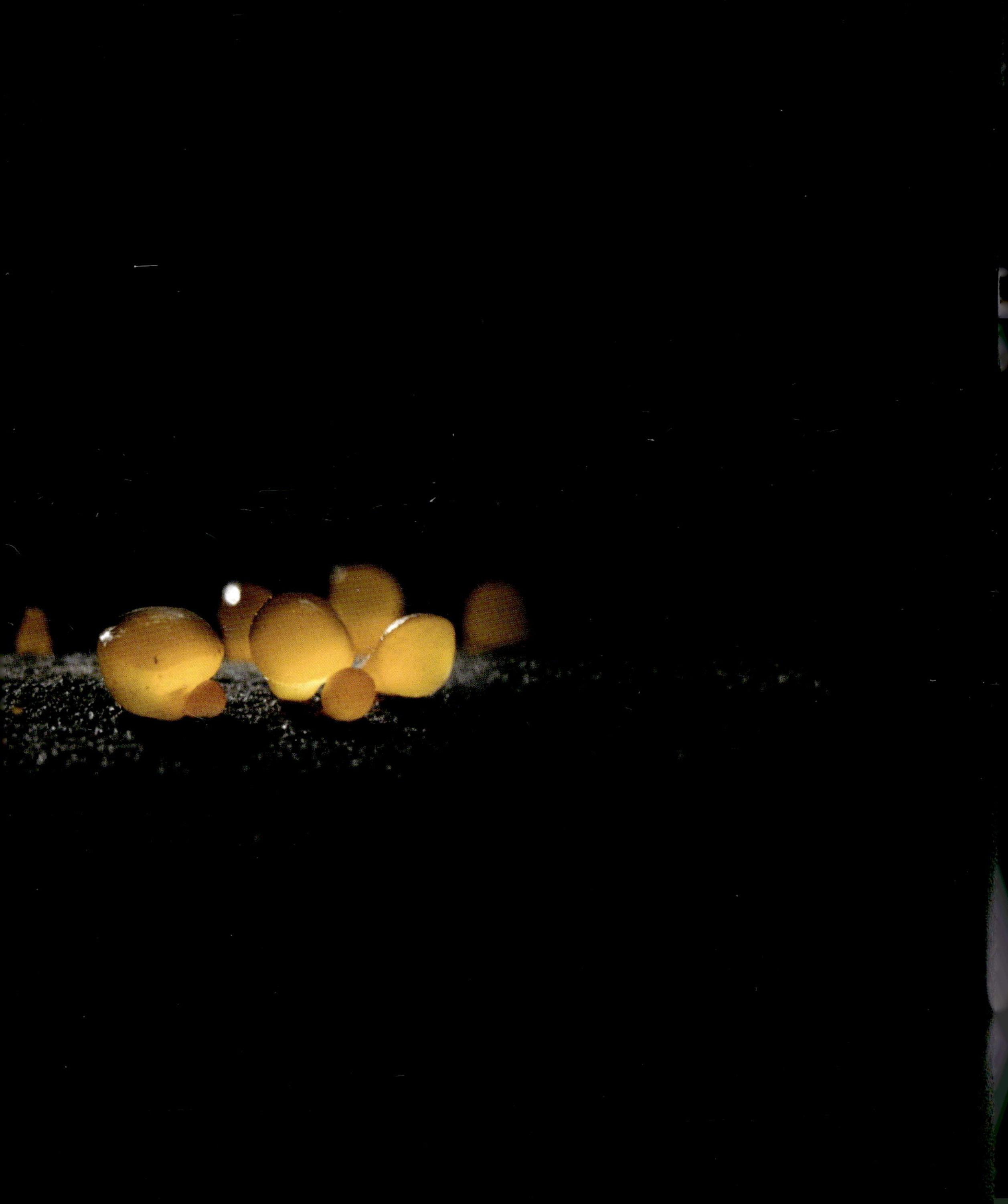

Funga Obscura

Photo journeys among fungi

Alison Pouliot

The University of Chicago Press
Chicago and London

Previous page:
Golden jelly bells
(*Heterotextus peziziformis*),
Australia

Following spread:
Smooth cage
(*Ileodictyon gracile*),
Australia

For Doreen Mai Hutton 1904–2005

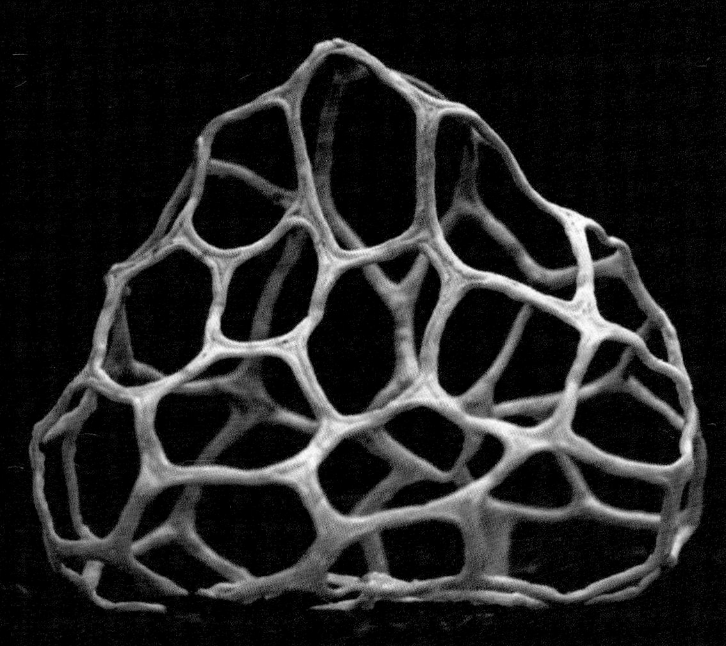

Focus on fungi

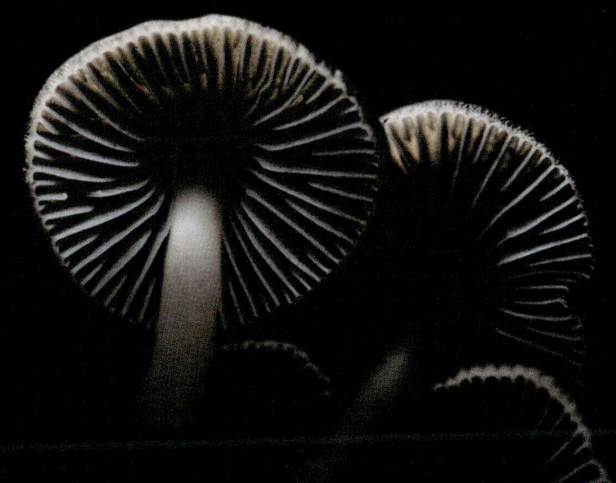

A wolf-whistle reverberated through the forest, followed by a cracking, ringing whoop. Then a gurgling croak. A shadow flickered through the shining gums in the slanting light. I startled as something landed on the forest floor beside me: a pied currawong. She steadied herself, threw back her head and continued her recital.

Dusk was approaching fast on the Errinundra Plateau in the south-eastern corner of the Australian mainland. I packed up my camera under the bird's yellow-eyed supervision. The bright beams of summer sunshine had softened as autumn beckoned and the sun sank low. Shadows lengthened, defining everything they touched. It's the trick of autumn as the Earth tilts on its axis, shifting spectrums of light, deepening colours. The glare turns golden. There's a gentle earthly exhalation, a post-summer sigh. A sense of fleeting, of letting go. A reminder of impermanence. I felt a growing sense of anticipation as another realm awakened.

I scanned the forest floor around me. I couldn't yet see fungi, but I could smell them. Lowering my nose to the ground, I inhaled a distinctive fungal funk. The currawong watched on. As if in imitation, she flicked her beak through the leaf litter while keeping me firmly in her gaze. My eyes were adjusting from the shapes of summer to those of autumn. I rolled over some bark and there they were! Gossamer threads of mycelia wended their way through the layers, weaving them together. Soon these furtive fungi would fashion their mycelia into mushrooms and heave their way through the earth.

The mellow and melodious song of a shrike-thrush pierced the cooling air. Everything around me was subtly moving, shifting, metamorphosing. Something brushed my face and landed on my nose – tiny spiderlings adrift on webs, kiting their way to new terrains. Then, at last, a welcome misty drizzle set in. If we look and listen, smell and feel beyond the familiar, the extraordinary forms and hues of another forest come into focus: the fungi.

Fungi had me hooked from a young age: first because of their aesthetics, and then I discovered what they do. The deeper I delved into their realm, the more astonishing they seemed. They mesmerised and enthralled me, got under my nails and colonised my mind, as I tried to fathom the character and cadence of their ways. Decades later, as I wandered through the bush, endless questions arose – questions which will help me piece together the compelling puzzles of Kingdom Fungi.

The autumn chill crept into my bones as I headed for the last light on the horizon. Earthly scents and the promise of mushrooms rose with every step. The currawong took flight and whooped and whistled her way through the darkening forest.

From the shadows

This book is about fungi and the photography of fungi. The title – *Funga Obscura* – unites the two. Camera obscura is an ancestor of the photographic camera. Taken literally, it means 'dark chamber'. The Kingdom Fungi is also a dark chamber in that so much about fungi remains obscured. The word *funga* refers to the fungi of a region, and the fungi in this book span hemispheres.

I present fungi as a visual journey, shining a light to guide you through this cryptic kingdom of enigmatic organisms. Although often hidden in plain sight, fungi grow almost everywhere. Some fungi produce mushrooms and other structures for dispersing spores. We notice them on our lawns and in forests and woodlands, but fungi grow in almost every habitat, from the ocean depths to deserts to alpine peaks. Yet this is not a book just about mushrooms. It goes beyond these spore vessels to explore how fungi shape landscapes and interact with other organisms, and it challenges our ideas about what fungi are, what they do and why they matter.

Our journey into the fungal kingdom starts long ago in evolutionary time, when fungi emerged from the ocean onto land. We begin in elemental landscapes of ice and rock, then follow the path of fungi as creators of soil, enablers of life on land, players in the succession of forests and recyclers – including of things humans create. We traverse time and space, crossing continents and ecosystems, to witness their work. It's a visual celebration of the remarkable lives of fungi, to capture their essence within their habitats and among other life forms. You might recognise a few of them. Sometimes their identities are apparent; at other times, there are just clues. My approach is poetic rather than diagnostic, with the images leading the way and telling the story, or a story that you might discover yourself. There are many ways to know fungi. This is just one. There are also some essays that reflect on the visual story.

For me, being in the forest is to presume a particular relationship with fungi: raw, unmediated and in their terrain. Images of the natural world, especially visualisations of scientific data, often give us a detached, macro-scale perspective of planet Earth. Rather than an all-encompassing, top-down approach, I want to take you with me, to the forest floor, to discover fungi up close.

To understand fungi, it helps to tease out their interplay with other forest inhabitants, the eternal kinetics of weather and seasons, the soils and their histories, and all the various goings-on that shape fungus lives. As artful opportunists, fungi are supreme colonisers.

Fungi don't grow *in* these environments per se; they help *create* the ecosystems that sustain the planet. Through their dealings with trees and other organisms, and their unlocking of nutrients, fungi do the groundwork. These transformative synergies, exchanges and processes are what make a forest a forest.

Fungi are very much in the Zeitgeist. Some people seek them as a source of food or enlightenment; others harness their medicinal or pharmacological properties. Mycologists reveal their evolutionary histories and relationships, and assign them names. The superstitious wielded them to ward off witches or to contact the dead. Fungus mycelium has been warped and wrangled to create furniture, underwear and coffins. Fungi have become symbols of hope for restoring a damaged planet and for imagining a different future. Yet they occupy the uneasy space between the fungally fearful and those who hope they'll save the world. We are enthralled by the vital exuberance of fungi, and at the same time unnerved by their habit of recycling the dead. Fungi wobble the assumptions and frameworks we use to understand the natural world. They trouble the order of things, operate counter to what we perceive as normal, and mess with our systems for categorising organisms. Their strangeness and the enigmas they conceal add to their appeal.

Through the viewfinder

I've got to know fungi mostly through direct observation and experience in the ever-changing conditions of the field. It's through the viewfinder of my camera that their astonishing lives have come most sharply into focus.

There are few things like mushrooms to put you in the moment. Mushrooms, and the matter that fungi deconstruct, are evanescent, and they make the present more acute. There's a tantalising temporal tension between capturing fungi photographically and their ephemeral lives. Fungi tempt us to fossick beneath the surface, both literally and figuratively. Each depression of the shutter requires a choice about how much to reveal or conceal, and whether to convey the science or poetics of a fungus. Every image is subjective. I make active choices about which fungus to photograph, whether it is representative of a species or an outlier, and if I'll reveal its identity at all. I select what to show and what to obscure, what to emphasise and what to understate, and whether to create a literal or abstract representation. I determine which moment in the mushroom's existence I want to convey and how to contextualise it. Fungus as abstraction? Fungus as specimen? Fungus as habitat? Fungus as anonymous? Fungus as identifiable?

Photographing fungi showcases overlooked life forms and challenges judgments about which organisms are deserving of our attention or admiration. Glamorous orchids, birds and other charismatic creatures have held the spotlight. Yet the diverse and peculiar forms of fungi upend notions of beauty and aesthetics and might inspire us to ask how such curious configurations come to be.

Images can expand our notion of what's worthy of our attention. Fungus photography thrives as we push the bounds of what might be considered credible subjects. They reflect the enduring capacity of photography to transform overlooked themes into meaningful ones. Yet historically, fungi have mainly been photographed for the limited purpose of identification. Most published images of fungi are taxonomic. I've created many of these too, but I'm more interested in how we might move beyond the limited scope of fungus image as scientific record or the melancholic familiar, and how we might spark more speculative possibilities. Could a mushroom photograph serve as a prompt to the future, rather than something fixed in the past? It's a challenge to bring visual drama or theatricality to a mushroom

Following spread:
Rainforest horsehair (*Marasmius crinis-equi*), Australia

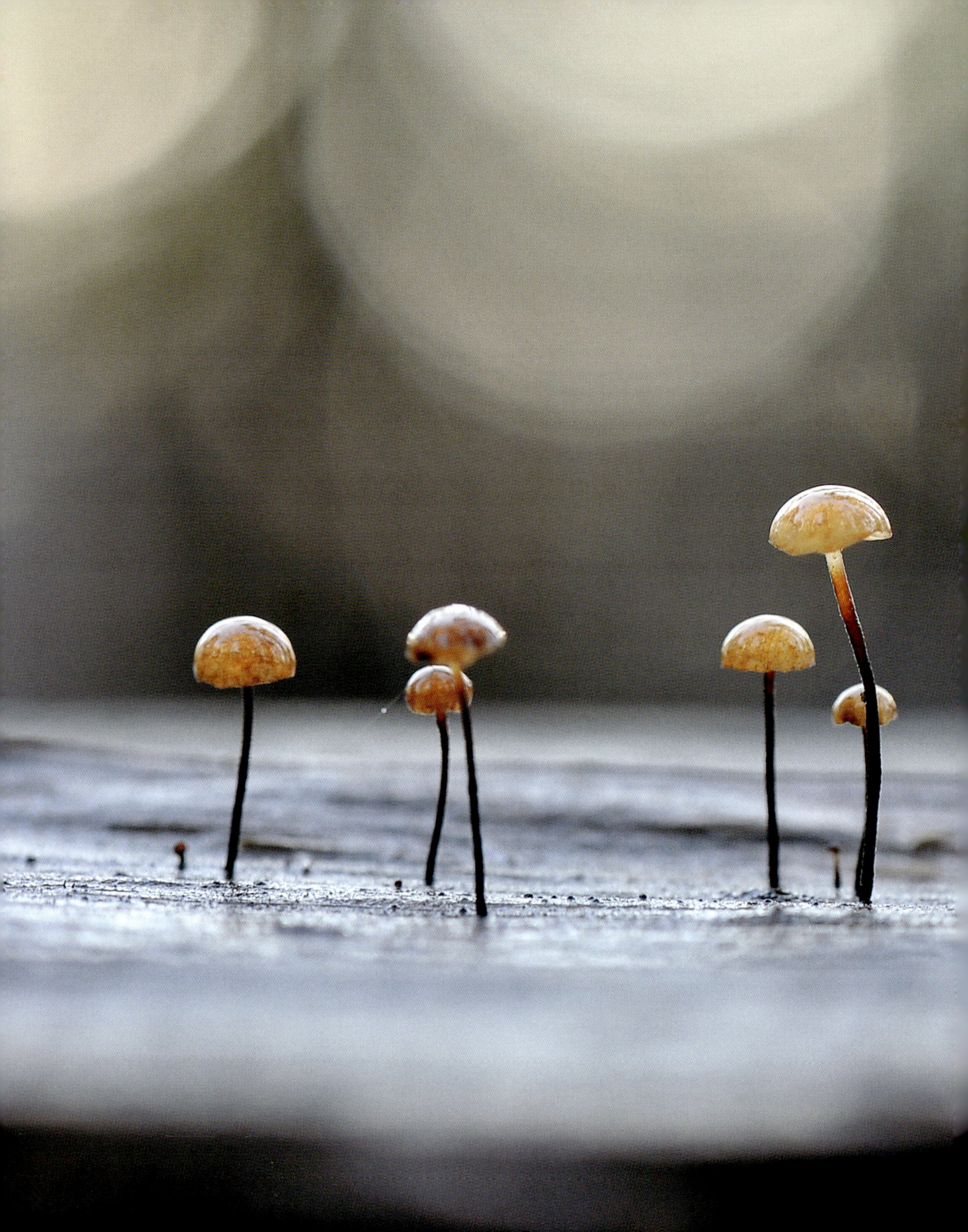

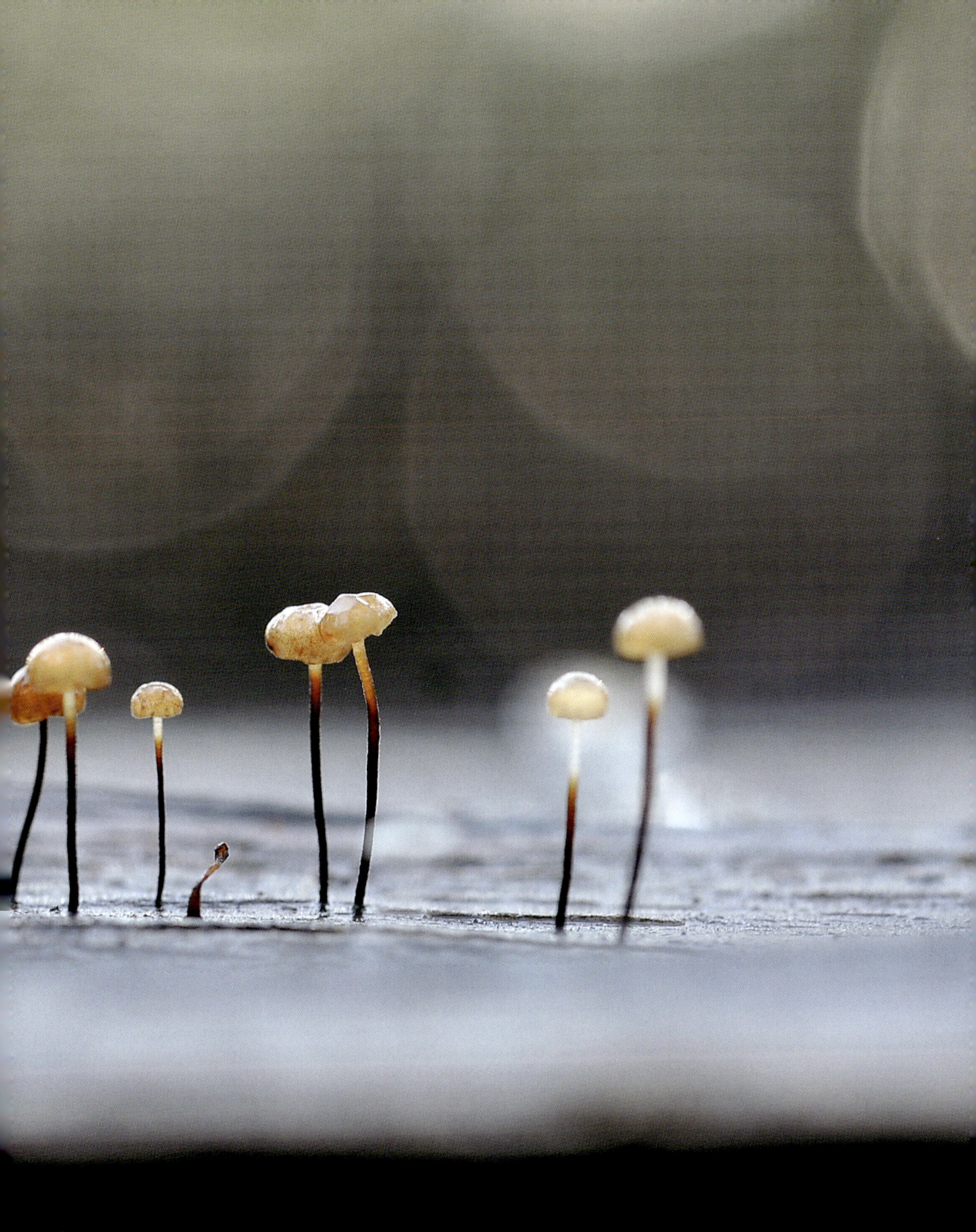

photograph relative to photographing, say, a human subject. Yet beyond being data and a static object, a fungus photograph can serve as an intermediary – a means to imagine beyond the frame, spur new modes of enquiry and representation and inspire a reset of our relationship with nature. As fungi become familiar through photography, these images not only open our eyes to the ecological significance of fungi but also shape our emotional responses to these organisms.

With the camera extending the senses, photographing fungi reveals each species in its many guises. Fungi are shapeshifters, appearing differently depending on the nature of their patch. A mushroom growing deep in leaf litter might develop a longer stipe (stem) to propel its precious pileus (cap) of spores above the surface and release them to the wind, whereas the same species in shallower matter could well have a shorter stature. Just when you think you're familiar with a particular fungus, it dresses in a new outfit, mimics another species or defies being named. It's then you realise you've strayed into that exciting part of the forest that Lewis Carroll's gnat ponders: 'Further on, in the wood down there, they've got no names.'

Every time I enter the forest, I discover something new. While I can anticipate some of the fungi I'm likely to meet, there's always one that takes me by surprise. It might be a mushroom growing in an unusual place or an eccentric morphological manifestation, or perhaps a myco-parasite merging two species into one. Sometimes they're indistinguishable from their surrounds, until you get your eye in and develop the 'search image' to see them. Once you do, you wonder how many others have escaped your gaze. But it takes time. Not all mushrooms are shaped like umbrellas. Seeing is more than ocular. It grows from experience and expanded visual perception. Your brain might need to find the fungal frame to recognise unfamiliar forms. You are likely to pass them by until the search image crystallises in your mind, your eyes awaken and an unseen fungus form looms before you.

I photograph fungi in situ because I like being in the forest. To know them, I need to be there. In a studio, lighting can be controlled. However, I enjoy working in the field because of the constant changes, the unexpected shifts, and the ways the forest mixes colours and tones as it filters light through the canopy. Ambient light and the mood of the forest appeal to my aesthetic, despite the biting, probing creatures that are part of the deal of spending prolonged periods on the forest floor. It's in the field that you become sensitised to subtle changes in the surrounds: a slight shift in wind direction, the rising scents in the forest after rain or the surreptitious movements of creatures in the leaf litter. Over time, you attune to the sheer complexity and responsiveness of organisms commonly considered dull or inert. You get a sense that the forest is sensate and perceptive, alive and wriggling. Ever-present among fungi are other spineless lives. If you watch a mushroom long enough, a springtail is likely to whiz past your nose or a slug will cruise by, grazing on the mushroom's cap until it is a miniature minefield of holes. A snail will rapidly retract its stalked eyes if you move within its sphere. A spider will respond with a nervous crouch if you sneeze. Male flies will hold their ground on a mushroom, defending their space should you come too close. And there are no flies on jack jumper ants. When they detect you in their vicinity, they rear up on their hind legs, red antennae twitching.

Noticing the intricate lives of small creatures helps us switch scales and recalibrate to another level of existence. More often, larger and louder organisms grab our attention. In my Australian homeland, there are a lot of loud creatures. If you've ever been close to a cantankerous cockatoo or between a trio of laughing kookaburras, you'll know what I mean. *Homo sapiens* can be rowdy too. Thankfully, fungi are more quietly spoken. The great godfather of Italian gastronomy and fungus enthusiast Antonio Carluccio spoke of mushrooming as 'the quiet hunt'. The tranquillity and slowness of searching for mushrooms can be as much a part of the enjoyment as finding them. Does mycelium respond to my presence? It's hard to know because I can't detect that, but I'll keep asking the question.

Fungi and their friends are shifting status from largely unregarded organisms to interesting, important and perhaps even worthy allies. I photograph fungi not so much to name them, but to visualise the fragile and intimate relationship we have with all of nature, including fungi. It's an exciting time to be among mushrooms and mycophiles to share some impressions. Life on land involves a fungal backstory. And that's where we begin.

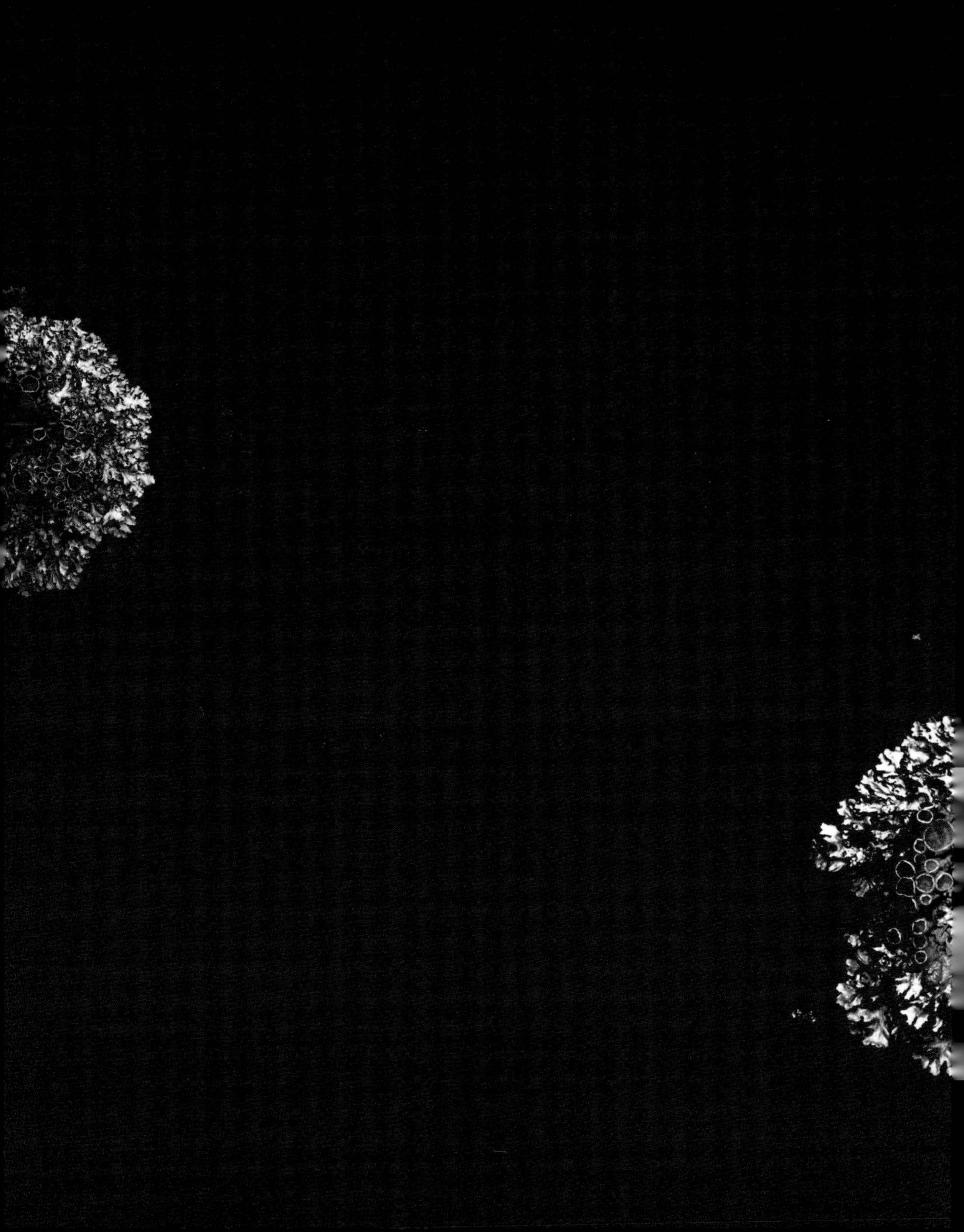

Pioneers

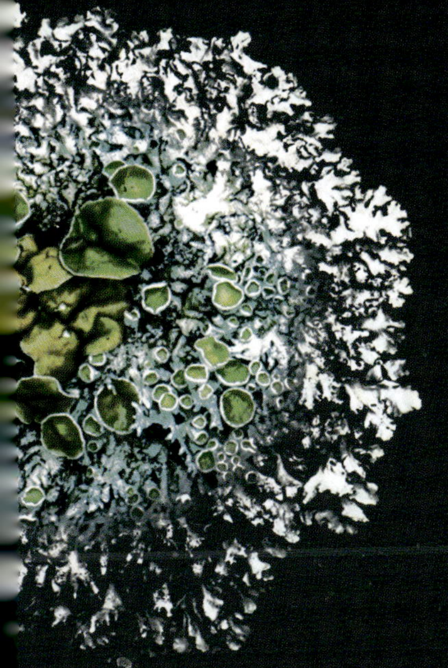

The punishing combination of bitter cold and biting wind makes it hard to linger long at 3610 metres above sea level on the summit of Üssers Barrhorn. Deep in south-west Switzerland's Pennine Alps, it's the highest point in Europe you can reach on two legs without ropes and crampons. My hiking companion and I had taken off from the Turtmann Hut well before dawn, traversing moraines and zigzagging up steep and slippery slopes of scree. As the sun rose, the dark, towering shapes of mountains turned from black to white and glacial blue. From the summit, the view of the 'four-thousander' elevations of Weisshorn and Bishorn opened out to an endless panorama of peaks. Scanning the spectacular 360-degree vista, my eyes smarted in the slicing gusts of wind as I wriggled my fingers in my gloves to try to retain some feeling.

It's hard to imagine a world that's purely elemental, a world without life. There are few such places on Earth where no life forms exist. Even when we can't see them, most environments teem with a frenzy of microscopic lives. But if you climb above the tree line, then higher – where the landscape is dominated by ice and rock and exposed to extremes of temperature, wind and solar radiation – you can envision how Earth might have looked before life emerged on land.

Ancient allies

A long, long time ago, probably a good half a billion years, if you can imagine (I can't), two organisms came together that hadn't met before. What happened next changed the course of life on Earth – albeit over another unimaginably long period of time.

Lichens are one of many different relationships that fungi form with other organisms. The lichen alliance involves a fungus and a photosynthetic partner (an alga or specialised blue-green alga called cyanobacterium). But that's not all. Single-celled fungi known as yeasts, along with bacteria and viruses, also live within the lichen entity. Making up most of the lichen's biomass, the fungus provides a protective body or thallus that the algal cells nestle within, tapping the sun's energy to assemble sugar. The yeasts' role is unclear, but they may help synthesise chemicals used in defence. It's a world-conquering, cross-kingdom conglomerate that has stood the test of (a very long) time.

Sharing talents and resources enables lichens to colonise bare rock and survive the harshest conditions imaginable, from the freezing poles to blistering deserts. Many lichens can endure prolonged periods of severe desiccation by entering a type of metabolic suspension called cryptobiosis, which brings their biochemical activity to a temporary standstill. They wait out tough times until conditions improve; then, with the first trace of moisture, they magically reanimate.

Today, lichens appear in diverse forms. Some adhere to rock, while others form a papery or powdery crust. The long, trailing tresses of beard lichens that drape tree branches can reach metres in length. Some lichens produce reproductive structures that branch or fork or look like miniature goblets. Others have lobes that are fringed or shaped like fingers or petals. Lichens possess a vast suite of chemicals and pigments that result in a vibrant palette of colours, from pink and carmine red to sulphur yellow, mandarin orange, and tones of greys and greens. And while we might mistake them for a stripe of paint or a bit of discoloured wood, if you inspect them with a magnifying glass, the complex tapestry of lichens colonising your garden path or letterbox might well astonish you.

I'm quite practised at spotting mushrooms on road verges from fast-moving vehicles (to the chagrin of any white-knuckled passengers), but I have also seen them from unexpected elevations. As I peered out of a plane window as we descended through cloud over Finland's Oulu Airport late one autumn afternoon, lichens loomed from the landscape below. Given they stand only a few centimetres tall, you'd expect to need an eagle's superior vision to see them. Yet the vast extent of reindeer lichens and their contrasting paleness among the dark-green tones of young conifers makes them conspicuous, even from altitude.

Finland is the most forested country in Europe and ground-colonising lichens are a vital part of the nutrient webs of its forests. Lichens form thick mats that retain moisture on the forest floor, sheltering soils and tree roots against extremes. They provide microcosms for invertebrates and a soft, insulative lining for the nests of numerous birds and mammals like squirrels and dormice. In winter, when food is scarce, lichens provide nutrition for hungry reindeer. As well as ground-dwelling lichens, there are many that grow on trees. More than 150 lichens and allied fungi grow on spruce alone. Some attach to tree trunks, others clothe branches, and further intrepid types prefer cones.

Lichens cover about 7 per cent of the Earth's surface. In alpine environments, they are often the only sign of life. Excelling at clinging becomes a matter of survival when exposed to severe gales and the inhospitable conditions of high-altitude environments. Lichens adhere to rock in different ways. Many have minuscule fungal filaments called rhizoids that serve as anchoring roots. Unlike the roots of plants, rhizoids are just for fastening and have no vascular capacity, so they don't supply water or nutrients to the lichen. Instead, lichens rely on being able to extract whatever skerricks of moisture and food they can from wafts of passing mist or drops of rain. Rather than rhizoids, other lichens have a holdfast or central peg that extends the lichen thallus, securing them to the rock.

Being among the first colonisers of rock when glaciers retreat, lichens – along with other free-living microscopic fungi – initiate the slow process of mineral soil formation, or pedogenesis. Lichens produce a potent cocktail of organic chemicals, particularly oxalic acid, and through the processes of chemical and physical weathering, they slowly erode, disintegrate and dissolve rock. When lichens die and decompose, they themselves contribute to these newly forming soils.

Lichens not only create soils but stabilise them, especially in arid areas. Along with other fungi, algae, bryophytes and cyanobacteria, lichens are part of the intimate association of biological or cryptogamic crusts that colonise the upper millimetres of desert soils. Various anchoring devices – rhizoids and other tiny filaments – penetrate and bind sand and soil particles into aggregates, stabilising the surface. The changes they create in the surface microtopography capture moisture, dust and organic matter, increasing the fertility of the soil and its capacity to infiltrate and hold water.

Benevolent bryophytes

As lichens kickstart the process of pedogenesis, they create more favourable conditions for the succession of plants that follow. Among the early plant pioneers are the bryophytes – mosses, liverworts and hornworts.

Life on land presented many challenges for the first organisms that transitioned from a marine to a terrestrial existence. Like most non-vascular plants – those without roots or a plumbing system of vascular tissue – bryophytes overcome the issue of gravity simply by being small, usually only a few centimetres tall. By hugging the rock surface and being lightweight, they reduce the risk of strong winds blowing them away.

Because they are small, bryophytes are overlooked. Larger plants, especially the showier flowering plants, are more likely to grab our attention. But if you lie on the ground, come down to their level and reduce your focus, it can feel like you're in a miniature primeval forest. Masses of moss sporophytes, like minute microphones or seas of antennae, suspend sparkling droplets of water. They do so with a micro-architecture of water-catching grooves, bumps, wrinkles, hairs and hooks. Their tiny leaves and other absorptive surfaces act like sponges, capturing and holding water. Giant creatures clamber and roam among them. They're really just mites and their kin, but up this close they're intimidating.

Like some lichens, mosses have rhizoids rather than roots that anchor them to the rock. While some mosses can slurp up nutrients through their rhizoids, most absorb them directly from the moisture around them. They continue the groundwork of lichens in the early stages of ecological succession, stabilising soil surfaces, reducing water evaporation and making these terrains more habitable for subsequent colonisers. During a rainstorm, they cushion the energy of raindrops, diminishing the scouring effect and curtailing erosion. When the rainstorm passes, mosses slowly release the water they have retained into the soil.

Heading skywards

Descending onto the Rhone floodplain in the réserve naturelle du Vallon du Longet, where the river marks the French-Swiss border and 'living fossils' or relicts still hold sway, is like travelling back in time. Horsetails are the sole surviving genus of an ancient lineage of plants that were abundant in peat forests of the Carboniferous era. Along with ferns and lycopods (club mosses, spike mosses and quillworts), horsetails are vascular plants with roots, stems and leaves but, like fungi and bryophytes, they reproduce by spores, not by seeds. They probably survived the radical changes over the aeons by sheltering in rare ecological refugia – tiny remnants of what were once more expansive ecosystems. Today, the dinosaurs of that time are long gone, but mushroom foragers skulk through the undergrowth, and some are every bit as fierce.

Horsetails provide a genetic link back to a time when, 30 metres tall and towering over ferns and bryophytes, they formed the first ancient forests. Today, horsetails have reduced in stature and are rather spindly, flimsy things, yet they are stunningly beautiful when light filters through their feathery fronds. Their delicacy belies their hardiness as supreme colonisers that flourish in disturbed environments. Horsetails' success hinges on their flexibility to reproduce sexually by spores and asexually by deep-rooting rhizomes that allow them to colonise and rapidly spread.

Vascular plants evolved roots made of vascular tissues (including xylem and phloem), which, compared to rhizoids, are not only better at anchoring a plant to the ground but are much more effective at absorbing and transporting nutrients and water. The biggest advantage they gained over the bryophytes was their ingenious invention of a compound called lignin. Lignin gave them rigidity and structural support, allowing them to stay upright even in times of water shortage. Having an internal plumbing system radically changed the way these plants operated, enabling them to move away from swamps and over the land, accelerate soil development and create more complex forests. Being made of vascular tissue and lignin, they could grow much taller and up into the light. While bryophytes pioneered the land, early vascular plants took to the air.

However, most organisms cannot digest lignin. And here's where fungi step in. Only fungi can degrade lignin, breaking down dead trees and recycling nutrients, allowing forests to thrive and resulting in an explosion in fungal diversity.

The jury is still out as to the actual sequence in which the early terrestrial organisms appeared on Earth. It seems the evolutionary history of lichens might be shorter and more complex than scientists once thought. Identifying early lichens from a scanty fossil record is not easy, but lichenologists studying the evolutionary histories of lichens and plants recently concluded that early lichen fossils are younger than the oldest plant fossils – meaning that lichens may have appeared *after* bryophytes, ferns and other early vascular plants. Bacterial mats and mosses could have covered the bare rock 440 million years ago, with the lichens appearing later, around 250 million years ago. It's also likely that the lichen partnership developed several times in different regions. One thing is for sure: even 150 years or so after the Swiss scientist Simon Schwendener proposed his 'dual hypothesis' of the lichen alliance, and German mycologist Heinrich Anton de Bary coined the term 'symbiosis', the multi-partner lichen complex is slow to reveal its long-held secrets.

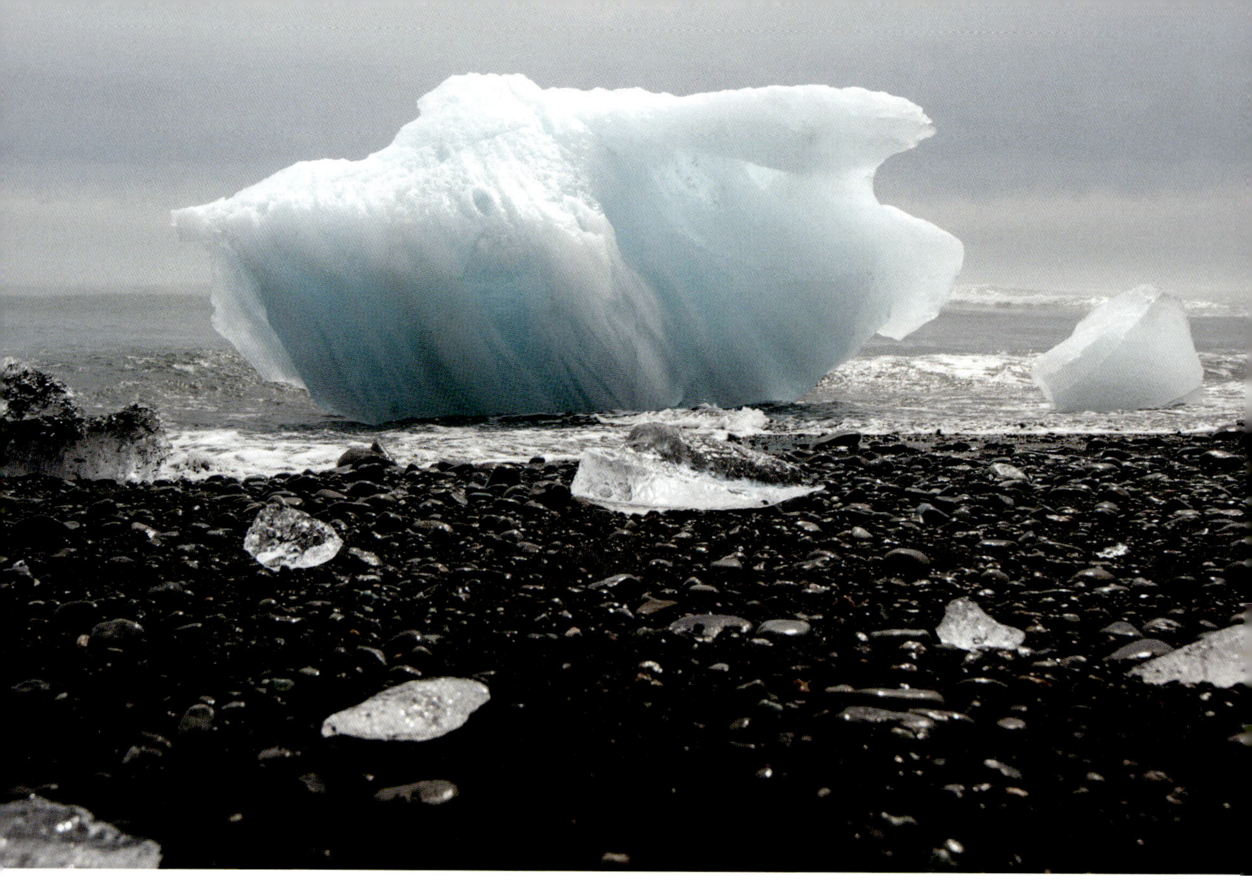

Elemental landscapes:
iceberg, south coast,
Iceland (opposite)
and Aletsch Glacier,
Switzerland (below)

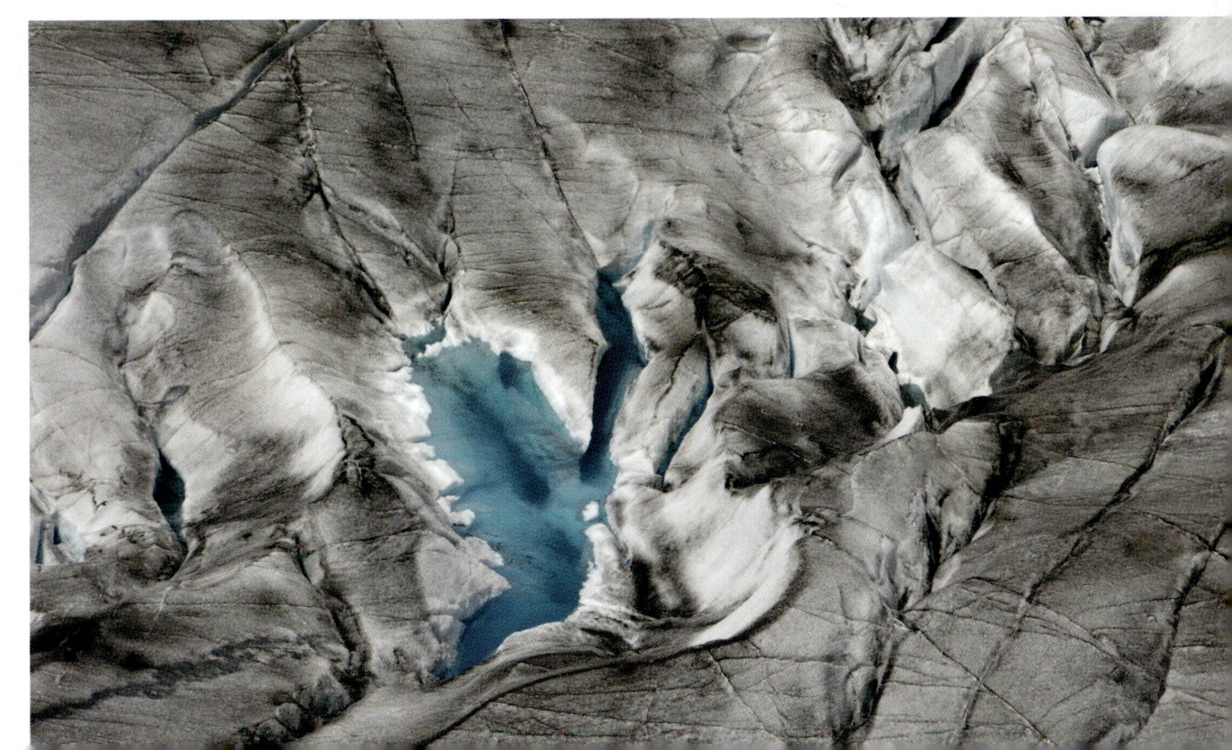

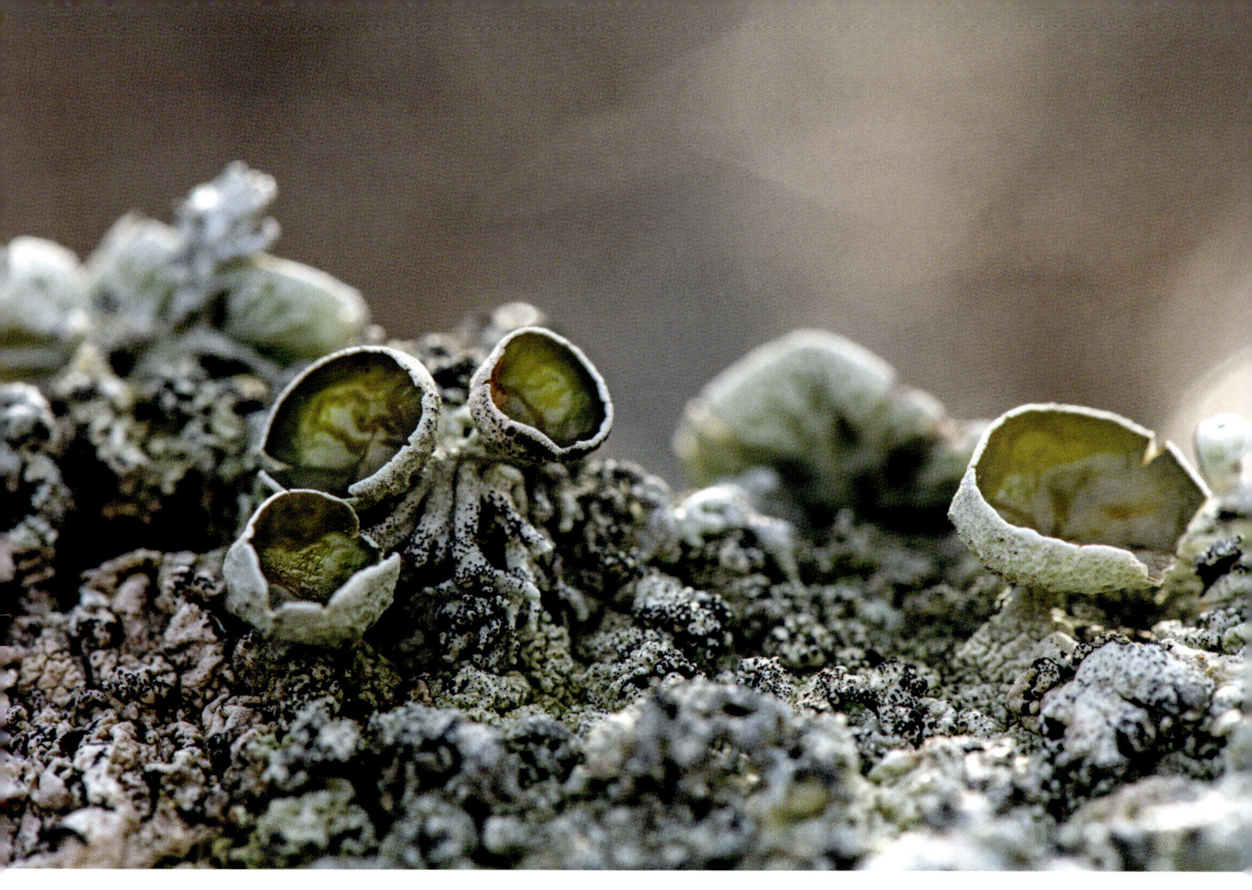

Lichen apothecia
(*Flavoparmelia rutidota*),
Australia

Crustose lichens, Australia (below)

Following spread: Crustose lichen community (*Pertusaria*, *Baeomyces*, *Amandinea*), Australia

pedogenesis

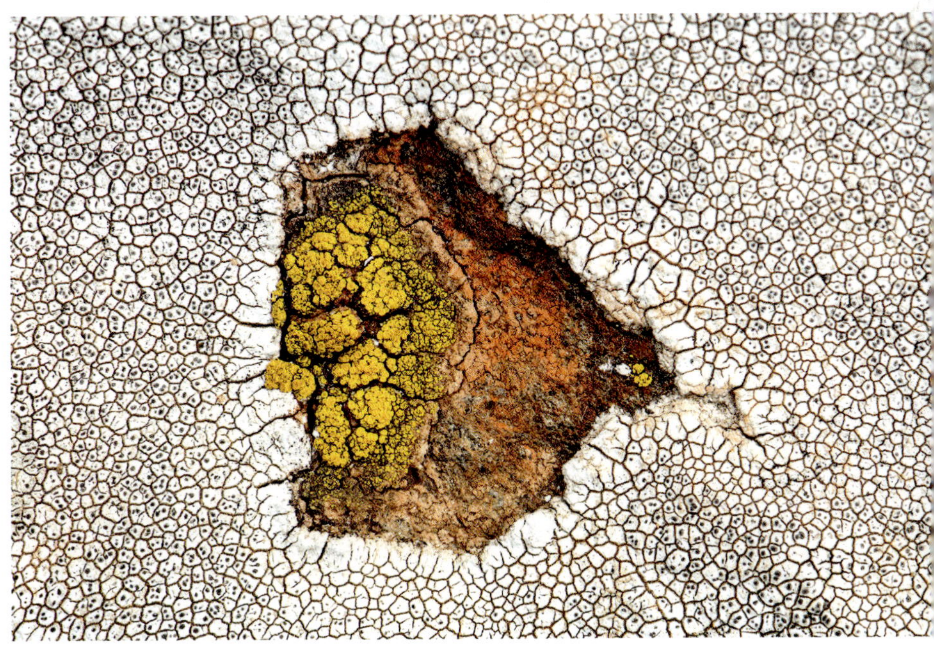

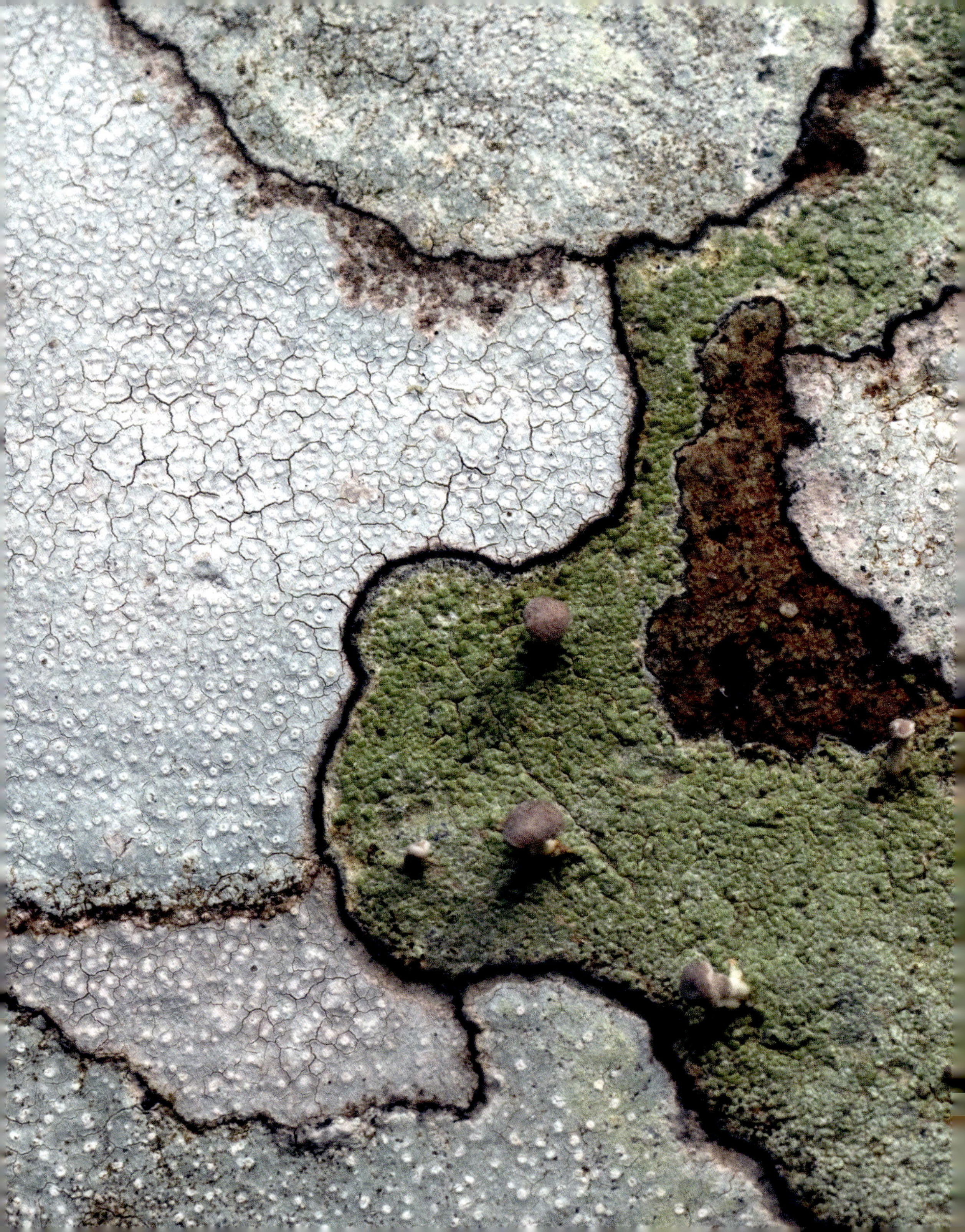

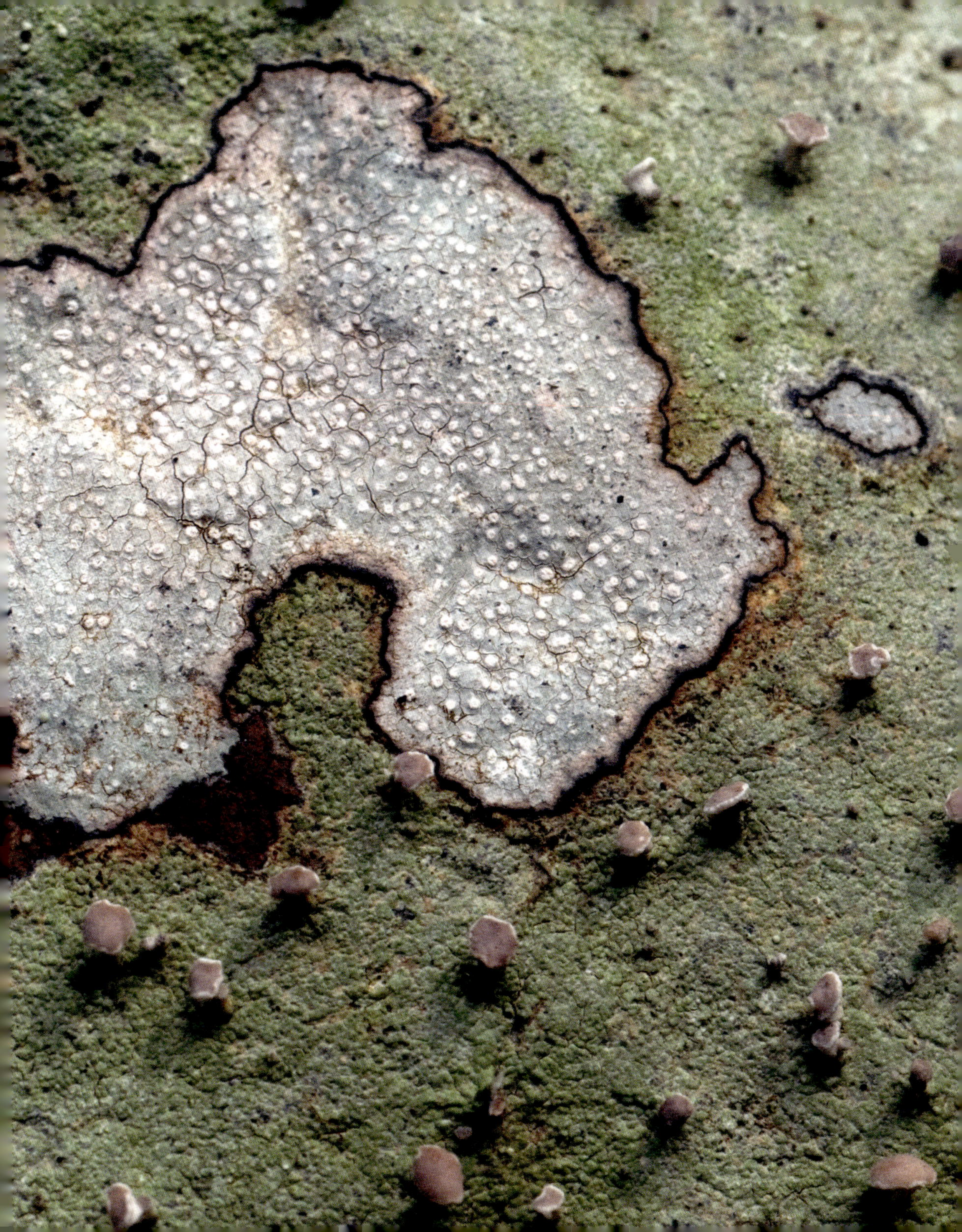

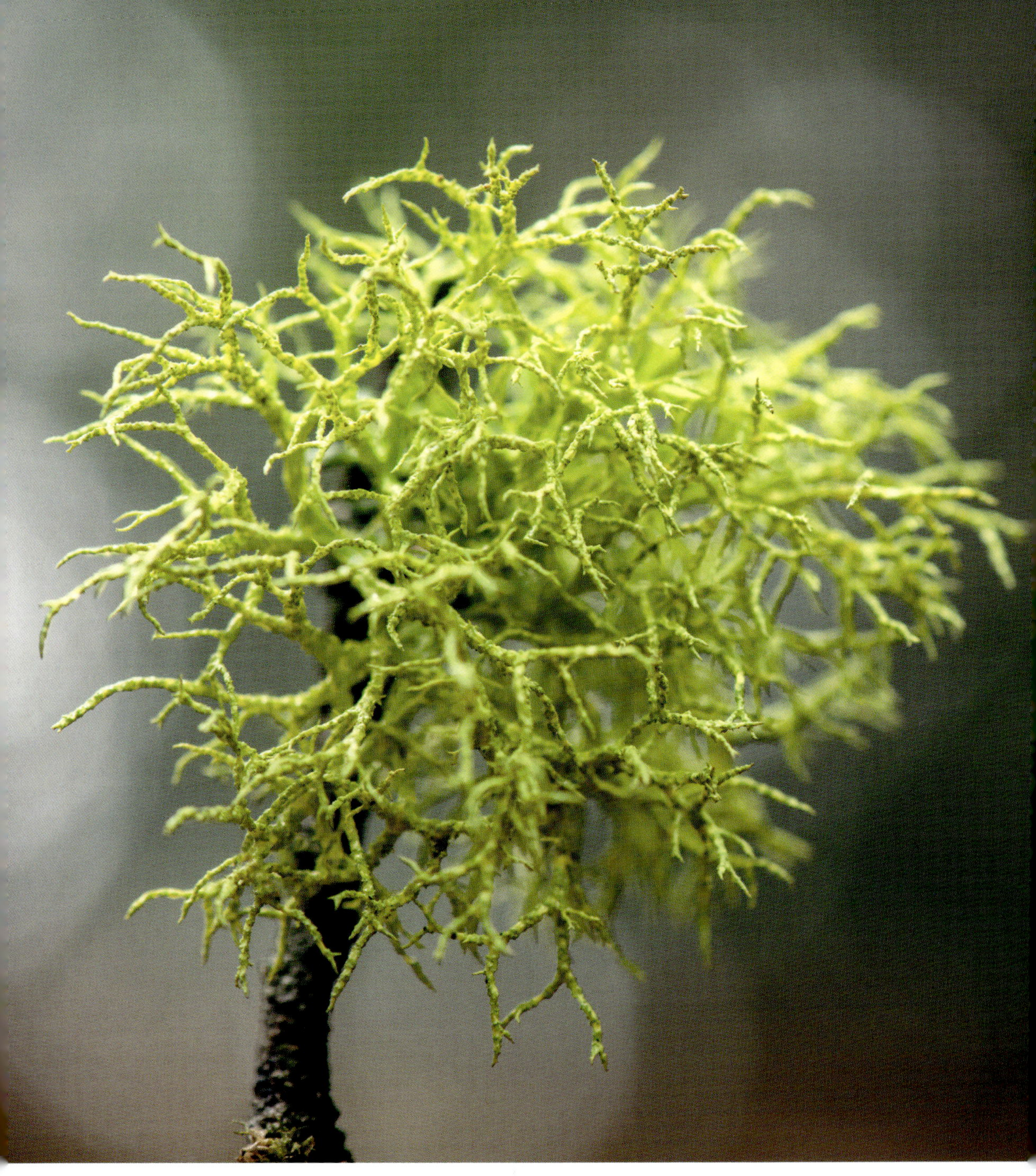

Wolf lichen (*Letharia vulpina*), USA

An alpine lichen
(*Pleopsidium* sp.), Italy

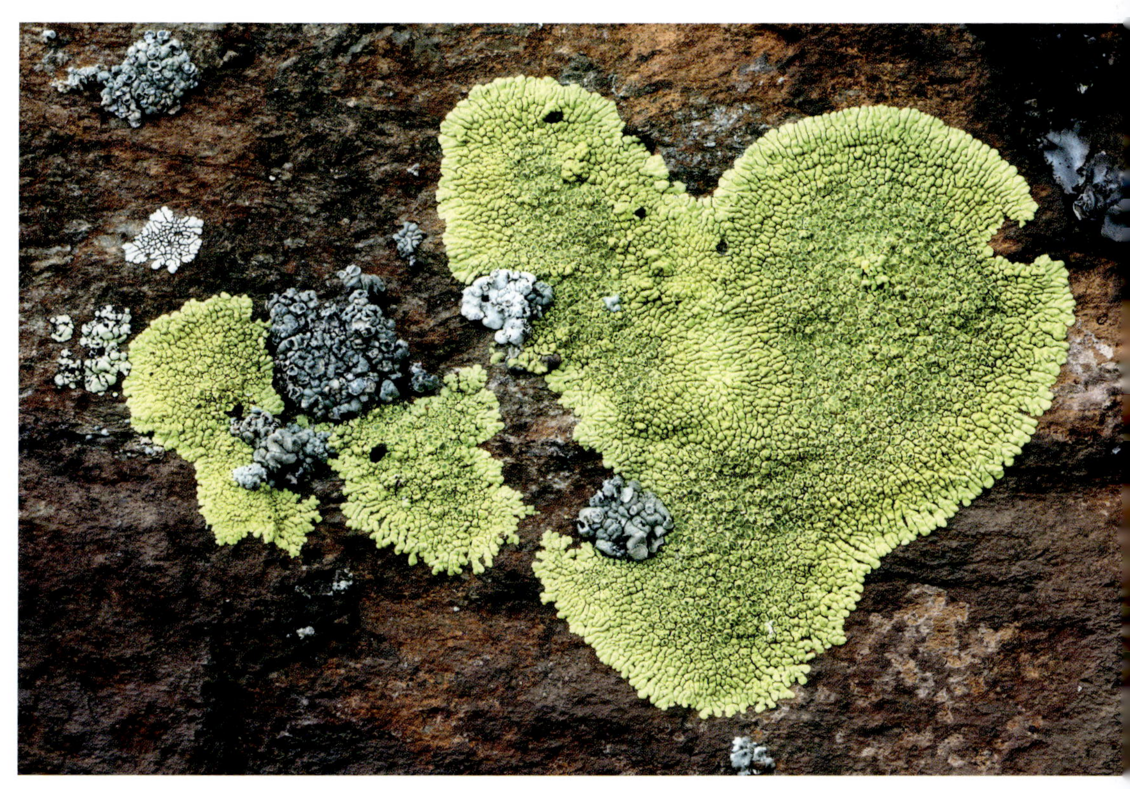

Opportunistic mosses seek out moisture on rock, Switzerland (opposite)

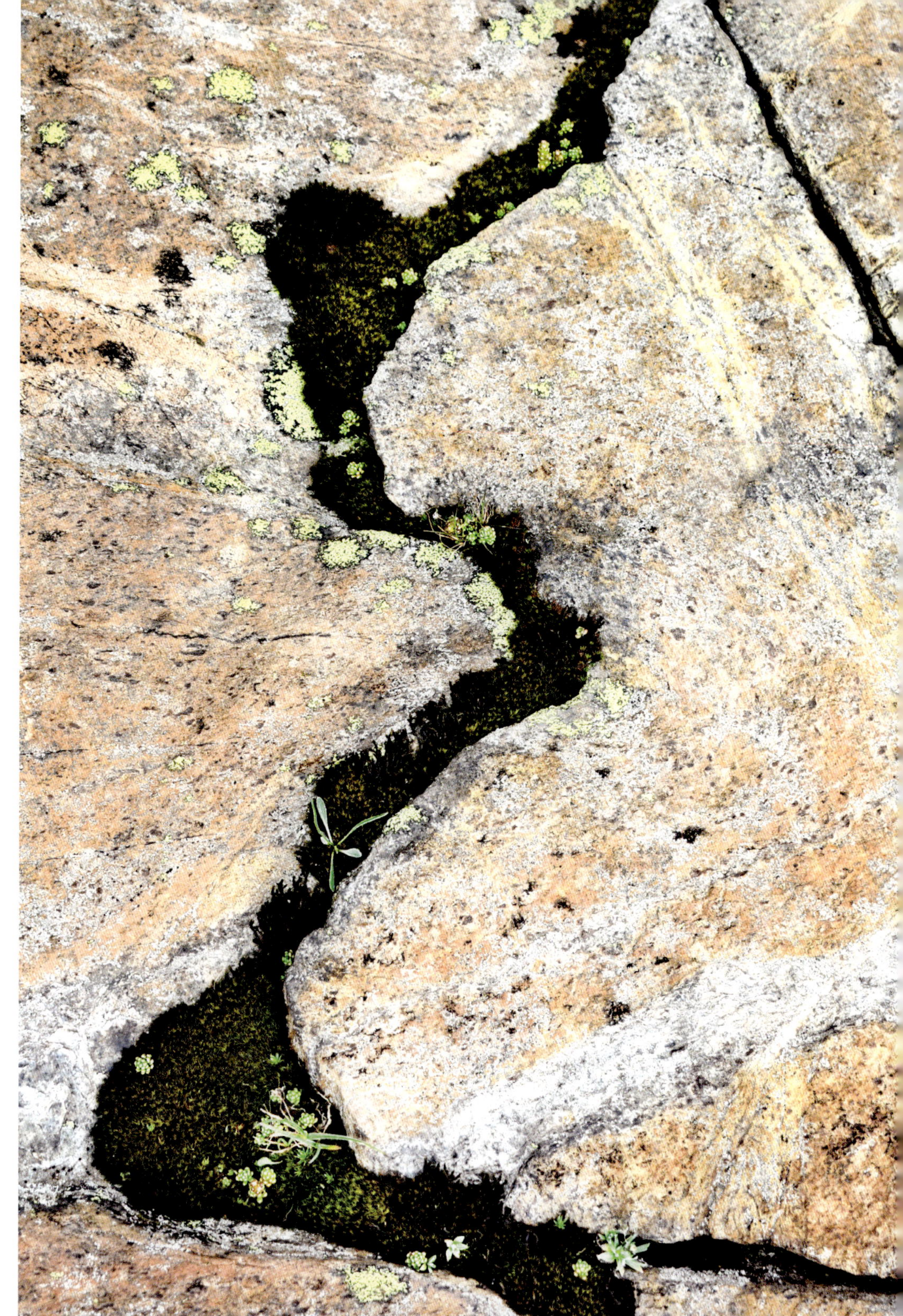

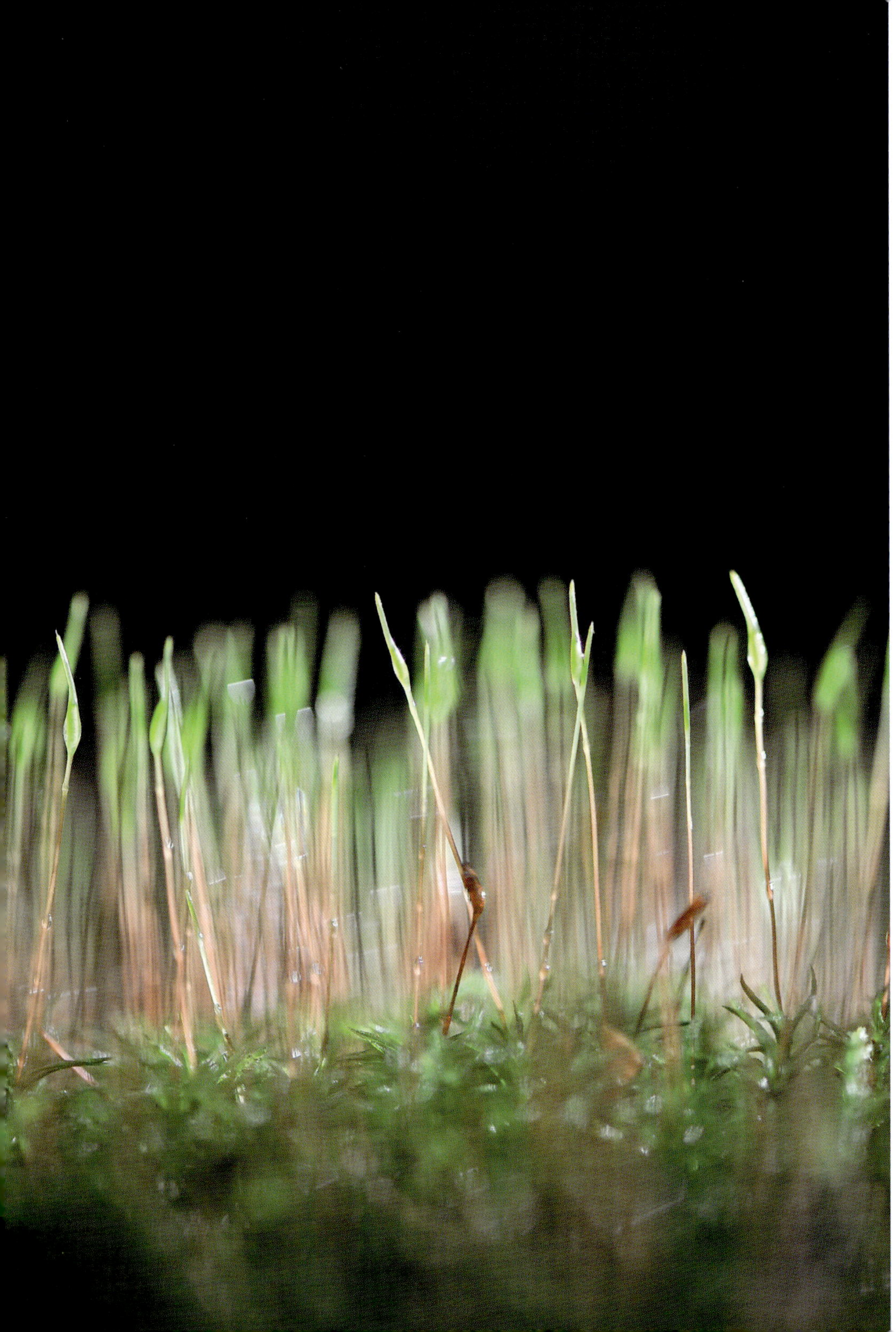

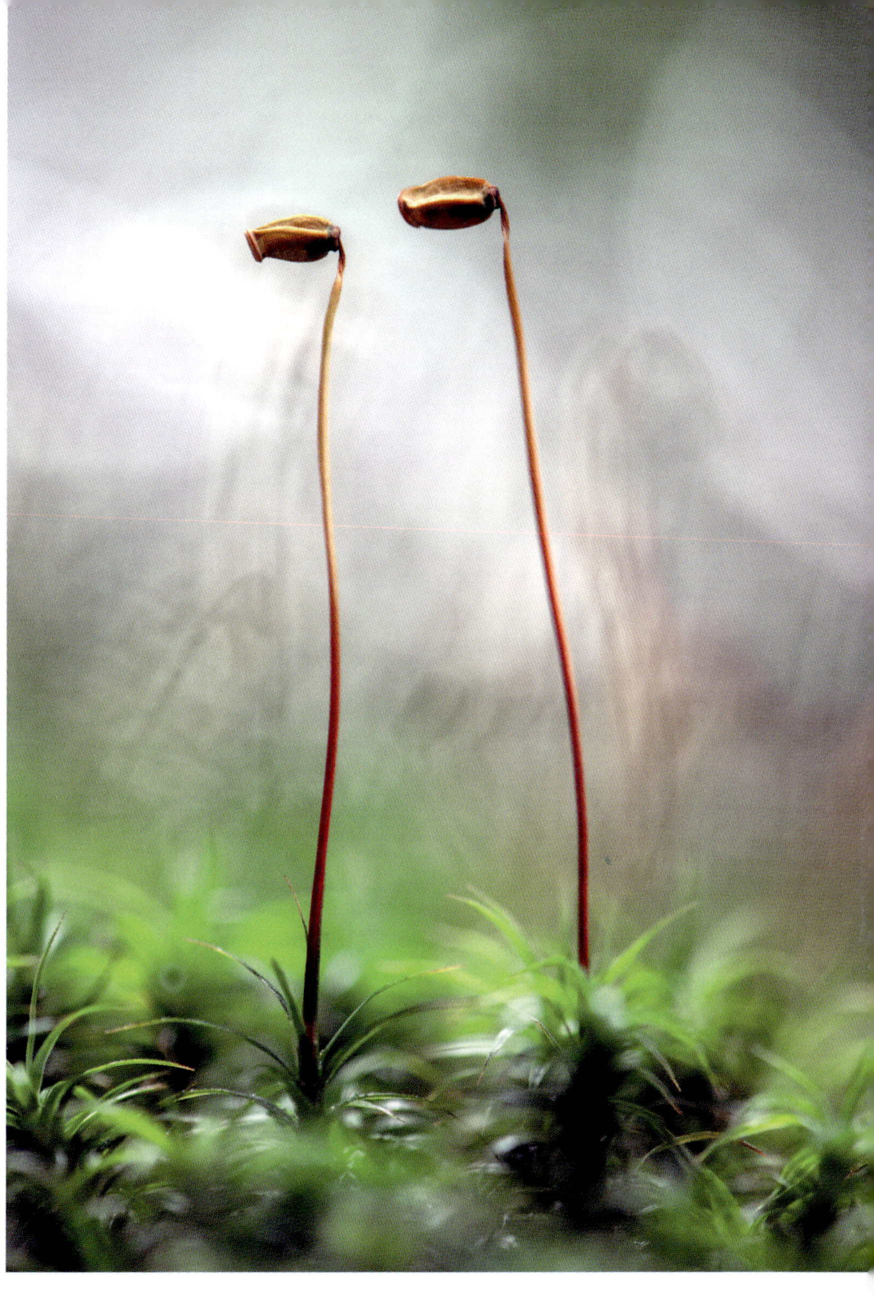

Juniper haircap moss (*Polytrichum juniperinum*), Réserve naturelle du Vallon du Longet, Switzerland (above)

Masses of redshank moss (*Ceratodon purpureus*) sporophytes, Norway (opposite)

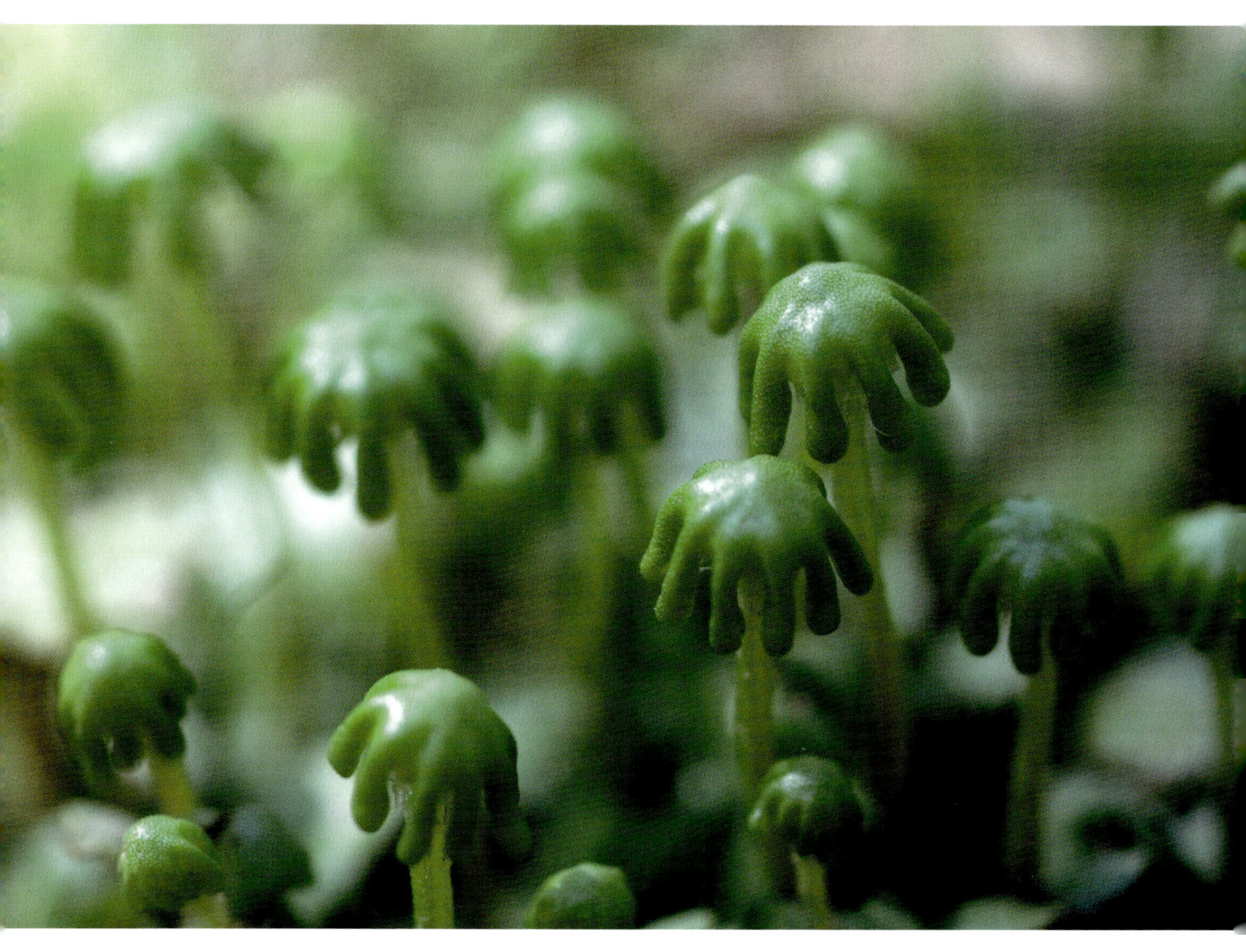

Bryophytes, like these umbrella liverworts (*Marchantia polymorpha*), are part of the intimate association of cryptogamic crusts

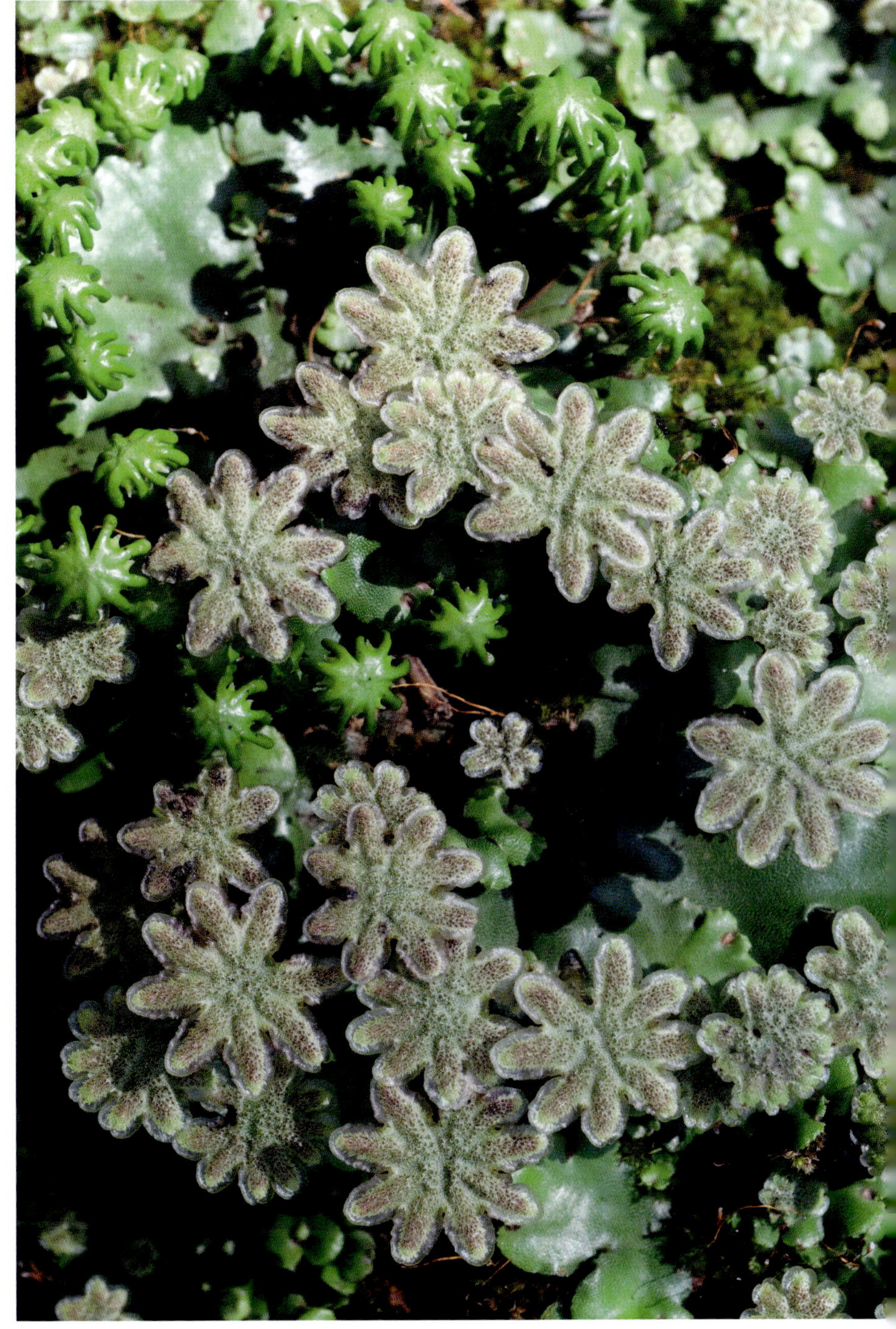

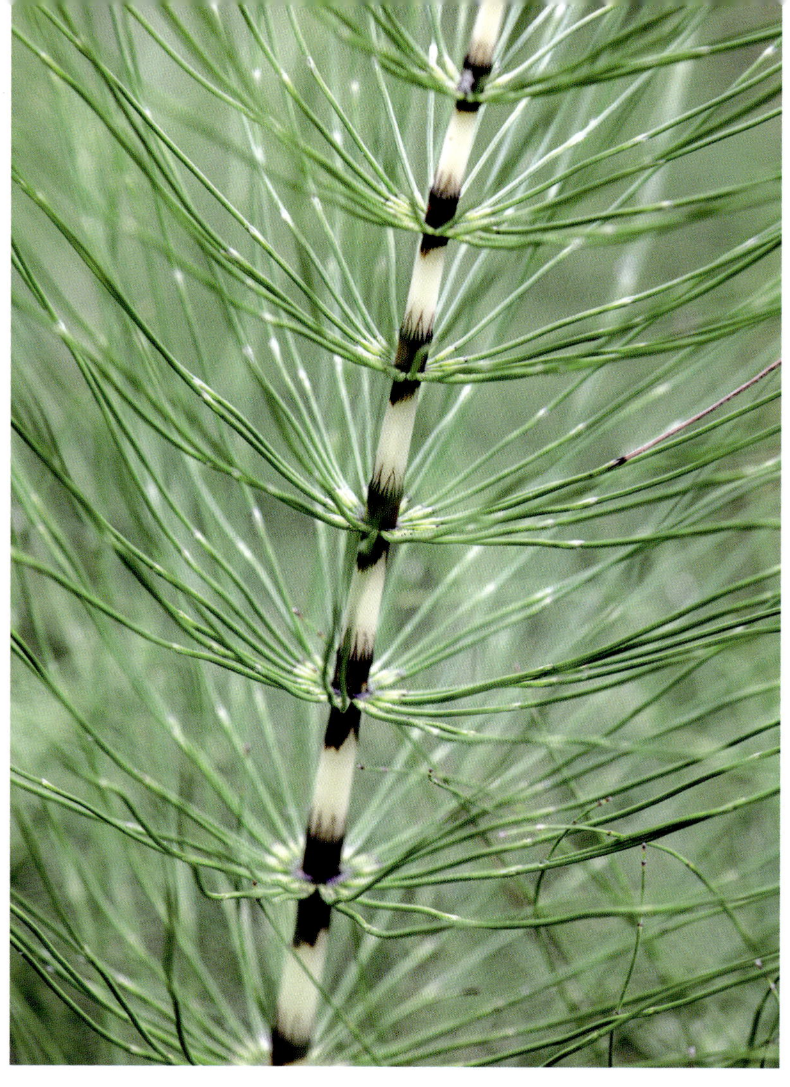

Horsetails, like this wood horsetail (*Equisetum sylvaticum*), are the sole surviving genus of an ancient lineage of plants

Fishbone fern
(*Blechnum nudum*),
Australia

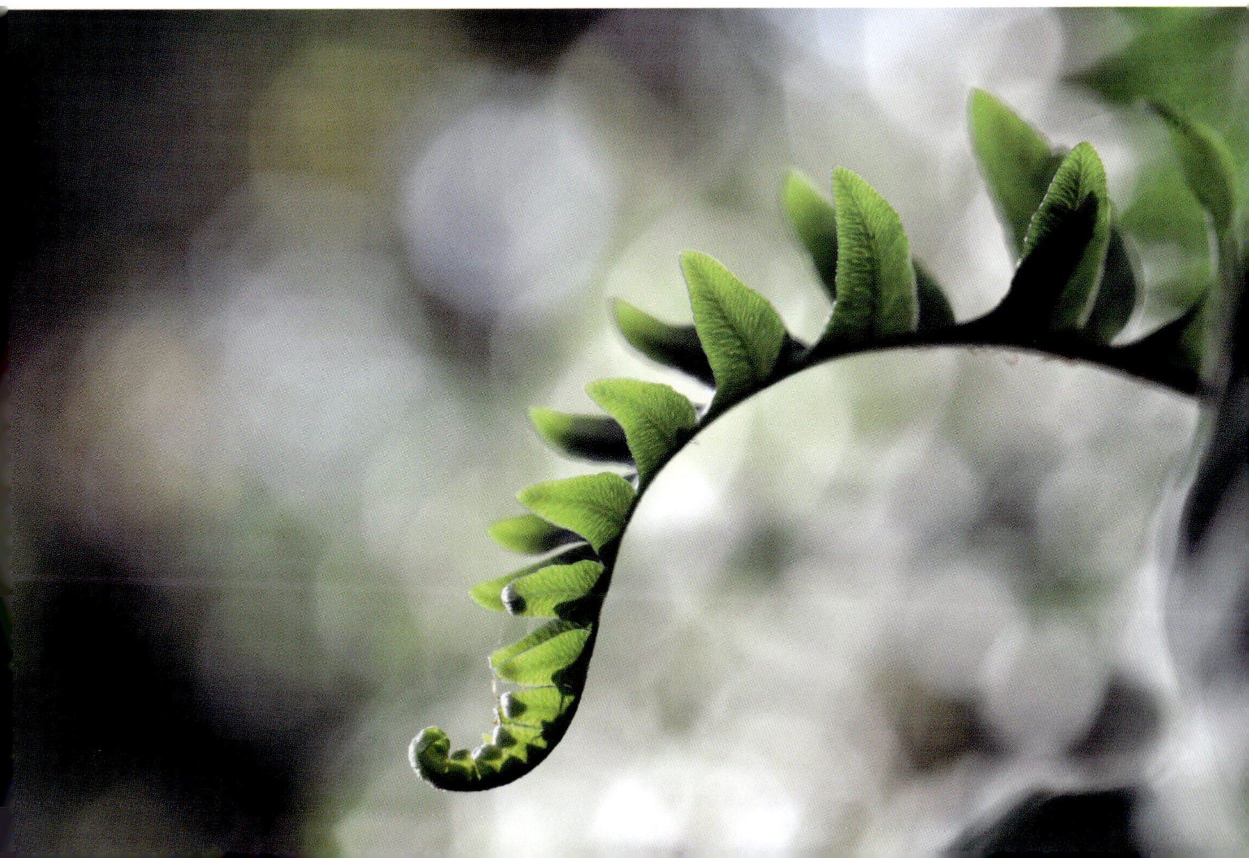

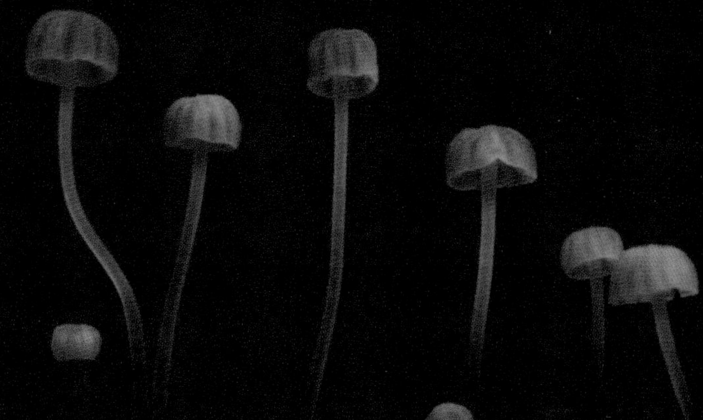

Into the forest

Looking upwards from the floor of the Laurisilva forests on the Portuguese island of Madeira, it's hard to find the sky. Some of these woody elders in the Parque Natural da Madeira are more than 800 years old and 40 metres high, and if you follow their gnarled trunks, the crowded canopy soon disappears into cloud. Along the way, you'll likely be distracted by branches draped with mosses, epiphytes and fungi. Further up, long thin strands of beard lichens gently waft in microcurrents of air and snatch moisture from mist. Tiers of shelf fungi nudge among the contorted conks of hoof fungi and clusters of translucent bonnet mushrooms. No surface is left bare.

Also known as cloud forests, they're the largest surviving relicts of once widespread laurel forests that clothed much of southern Europe between 15 and 40 million years ago. Mosses and liverworts upholster the twisted limbs and reclining trunks on the forest floor. These act as nurse logs, hosting and sheltering new growth as they themselves succumb and break down. They teem with fungi. Some poke out like antlers. Others form leathery goblets. Translucent jelly fungi cling to the undersides of fallen limbs; although jelly-like in consistency, they're surprisingly velvety to touch. Overlapping lobes of lungwort lichens – known indicators of old forests – fringe the margins.

Although mushrooms and other sporing bodies appear on the surface of trees and fallen trunks, it's within the wood that fungal hyphae are hard at work. Hyphae are much thinner than human hair and this allows them to penetrate pores and gaps in the wood that are inaccessible to other organisms. Fungi don't photosynthesise like plants; rather, they feed by dribbling digestive juices into the wood or substrates they inhabit. These potent enzymes break down complex compounds that the fungi then absorb as simple sugars. This way of gaining nutrition likens them more to animals than plants, but while animals digest food within their bodies, fungi do so externally. Most fungi gain their nutrition this way and have been doing so for a good half-billion years.

The age advantage

Like human societies, forests need elders. Old trees furnish a forest and its inhabitants in ways that young trees aren't yet able to accomplish. They're vital keystone structures that play unique ecological roles. Old trees are a buffer, or an insurance policy of sorts, that helps forests endure environmental change. As they age and become more varied in structure and form, they provide more overall area and a greater diversity of habitats. Cavities and hollows, cracks and fissures, develop with age, creating specialised habitats that are absent in younger trees. Beneath the soil surface, as roots hollow out at the base of an old tree, they allow creatures access to a catacomb of underground cavities and hideouts.

If you're lucky enough to walk in an old-growth forest, many things will strike you as different to a young forest. In managed forests, trees tend not only to be young but mostly of the same age, whereas old-growth forests have a varied mix of veteran trees, including those that are senescent and decaying, among a mosaic of young saplings – with every age in between. Trees of different ages and species all have different roles and functions that contribute to the dynamism and resilience of the whole. But perhaps the most obvious contrast is in the managed forest's lack of dead wood at various stages of decay, compared with a forest that is allowed to 'manage' itself in its own time. Primeval forests had vastly more wood than managed forests today. Prior to industrialised forestry in central Europe, for example, up to half of a forest's trees were old or dead wood in various forms – as stags (standing dead trees) and grounded logs, branches, sticks and twigs. There was even more in alpine regions, where decay and decomposition happen more slowly. The other conspicuous difference is that old forests are messy. And forests are supposed to be messy: trees seldom plant themselves in rows or chip their fallen limbs. Compared to managed forests, messy forests provide a broader range of microhabitats and microclimates that support a greater diversity of fungi.

In unmanaged boreal (high-latitude) forests, Scots pines (*Pinus sylvestris*), sometimes up to 500 years old, usually die while standing upright, and many remain standing for as long again. The Finns refer to dead, barkless, standing pines as kelo trees. When kelo trees fall to the forest floor, they provide niches for specialised fungi, particularly crust fungi (resupinate fungi) and polypores. Because kelo trees decay slowly, the fungi that live in them also take their time to colonise and reproduce. Hence their survival hinges on the continuous availability of these old trees, not just for years or decades but for centuries, even millennia. As their name suggests, crust fungi often appear as a simple 'crust' – a stripe or sheen, or a change in tone or hue of the wood. If you examine them up close, you'll notice their textures can be wrinkled or warted, wispy or cobwebby, pimpled, fuzzy, powdery or toothed. Some appear in extraordinary colours and strange configurations, with protruding shelves, rubbery lobes or radiating brain-like growths.

Afterlife

When a tree can no longer photosynthesise, although it's technically dead, it still has a rich and dynamic afterlife. Its structure provides a living, pulsing ecosystem, alive with an ever-changing cast of characters. Without the decay, disease and death brought about by fungi and other organisms, there is no life. The process of decay is productive and renewing. It's not an end, but a stage in a cycle of transformation.

Whether a dead tree is still standing or has fallen to the ground as a log, every part of it – including the trunk, bark and branches – provides a specialist habitat to myriad organisms. Over time, weather and nature sculpt its surfaces. Cracks and hollows form and capture the spores of fungi and bryophytes, and the seeds of flowering plants. Mosses trap moisture that increases humidity, alters local microclimates and accelerates decay. As fungi colonise and deconstruct the wood, they form further microhabitats for saproxylic invertebrates (those relying on wood) such as beetles, flies, millipedes and snails.

Dead trees constitute discrete ecosystems where the richness and composition of fungi changes over time. Many fungi are finicky, needing very specific conditions, and are hence highly sensitive to change. The size of a tree or log also determines which fungi inhabit it. A succession of organisms colonises an intricate microrelief of cracks and crevices, insect-bored galleries and chambers, and other micro-refuges. Each species further affects the conditions in some way and contributes to the process of decomposition.

Among the spineless inhabitants, beetles are numerous, with hundreds, if not thousands, of species occupying dead wood at different stages of decay. If you peek beneath the bark of conifers in the northern hemisphere, you're sure to spot early-colonising beetles like longhorns and spruce bark beetles as they scuttle into the deeper recesses. After a few years, as the bark loosens and the conditions within change, different beetles and other invertebrates move in. Various wild bees lay their broods in abandoned beetle tunnels in fallen logs, where their larvae feed on stored nectar and pollen. Mud dauber wasps and spider wasps seek uninhabited spaces to raise their families. Hover flies and crane flies don't eat the wood itself but the dead bodies and excrement of other creatures, while midge maggots graze on the fungi that colonise the log. As the wood decays, those who prefer their wood nicely rotted, along with armies of ants, stake out territory. With increasing humidity, the outermost layer of sapwood rots and an assortment of predatory invertebrates come to see who is for dinner. In the final stages of decay, with only the innermost supportive pillar of heartwood remaining, those that favour damp conditions – such as springtails, earwigs and earthworms – have their chance to take up residence.

Vertebrates also rely on dead wood habitats. The percussive tap of woodpeckers reminds us that they specialise in extracting invertebrates from within wood. They also hollow out homes in which to raise their young. In turn, other birds and mammals, including martins and bats, claim the abandoned woodpecker hollows. Frogs, toads, newts and salamanders make the most of pools of water or moister areas in the wood, while snakes and lizards seek snug spaces to curl up and hibernate over winter. In Australia's diverse woody habitats, these hollows are vital to parrots such as galahs, corellas and cockatoos, owls and treecreepers, many mammals, reptiles and amphibians and countless invertebrates.

Fungi take part in all stages of the wood-decay process, but many specialist fungi only occupy wood at late stages of decay. Therefore, breaking the continuum of old wood supply could result in the irreversible loss of these specialist fungi. The spores of most fungus species don't disperse very far, so it's not just about having a few old dead trees but having enough of them in proximity. If their woody habitats are fragmented or isolated, it's more difficult for them to disperse, and species diversity diminishes.

Forest archives

In Graubünden, Switzerland's wildest canton, the Cavaglia glacial mills at the foot of the Bernina massif record history shaped by the Palü Glacier. Flows of pressurised, gravel-carrying glacial water milled and eroded huge holes in the bedrock over millennia. Like the glacial mills, old-growth trees and logs are precious archives, and they too are older than the archives of measurements recorded by science. By recognising and reading the tracks and traces of fungi and company, you can tap into the histories of the endless comings and goings of species, the changes in seasons and climate, and infinite chronologies of events.

In the depths of the Poschiavo valley, I came across a huge old log, reclining on the forest floor like a crafted sculpture. I squatted down to examine it more closely and discovered hordes of fungi, mosses, liverworts and hidden creatures within its layers and convolutions. Yet this microcosm was not an isolated capsule of life. Rather, it existed in continuous interaction with the world beyond its woody borders. Constantly reshaped by the attritional processes of its inhabitants, it transformed and metamorphosed, accommodating new inhabitants as others disappeared. Secret messages in hieroglyphics were scrawled beneath its last fragments of bark – the tunnelling of beetles and their larvae, and signs of fungal disagreements. Mycelia don't exist in isolation, and when they meet, it's not always neighbourly. Zone lines demarcate the boundaries of different species (or different individuals of the same species). They are areas of interaction, or in the evocative German term, *Kampfgrenze*, or 'fighting borders'. You'd be forgiven for thinking that some creative person had taken to the wood with a fine-tipped black pen. The pigmentation results from the secretion of melanin, the same stuff that makes freckles. Woodworkers know this elaborate patterning as spalting and seek it as a decorative flourish.

You may have come across a piece of wood on the forest floor that is stained a beautiful shade of deep turquoise. Again, this is the work of fungi: in this instance, the delightfully named green elfcup or green stain fungus (*Chlorociboria aeruginascens*) that secretes a unique pigment called xylindein. The species epithet means 'becoming blue-green'. Sometimes you'll spot the tiny little cups of their sporing bodies on a log's surface. Another, *Phanerochaete sanguinea*, a rust-coloured fungus that grows on conifers, stains wood red. Or, if you gently pull back the bark on an old log or stump, you might see a latticework of the black rhizomorphs of honey fungi (*Armillaria*). Rhizomorphs are cord-like structures of parallel hyphae, and those of the honey fungus are encased in a melanised rind. This clever survival strategy of felting mycelium into robust ropey strands allows the fungus to traverse more territory, and transport water and nutrients over longer distances.

But these signs aren't always created by fungi. Many are the work of beetles and other creatures. Female beetles bore tunnels between a tree's bark and wood and lay their eggs. When they hatch, the larvae feed on the bark and drill their own tunnels that radiate through the wood. After they pupate, they eat their way out through the bark to discover that there's light at the end of the tunnel. Each species produces its own characteristic pattern of tracks. The egg galleries of western pine beetles have a serpentine pattern, while pine engraver beetles have a central nuptial chamber and egg gallery branches. Douglas fir beetles have vertical egg galleries with alternating groups of larval galleries, and others have horizontal egg galleries. Beneath all this industry you're likely to see little piles of excavated sawdust.

Further up the scree slopes, trees become sparse. The Arolla pine (*Pinus cembra*) is one of the few that can tolerate the harsh alpine conditions, where temperatures can drop as low as −40 degrees Celsius, and ozone and radiation become extreme. Its powerful roots wrap themselves around bare rock and penetrate deep into cracks to establish a hold and seek water. But they can't withstand these extremes on their own. They need their fungal partners to tap scarce water and nutrients. This is just one of many vital alliances that fungi share with other organisms, and these alliances are the secret to their success.

Following spread:
Laurisilva forest, Parque Natural da Madeira, Portugal

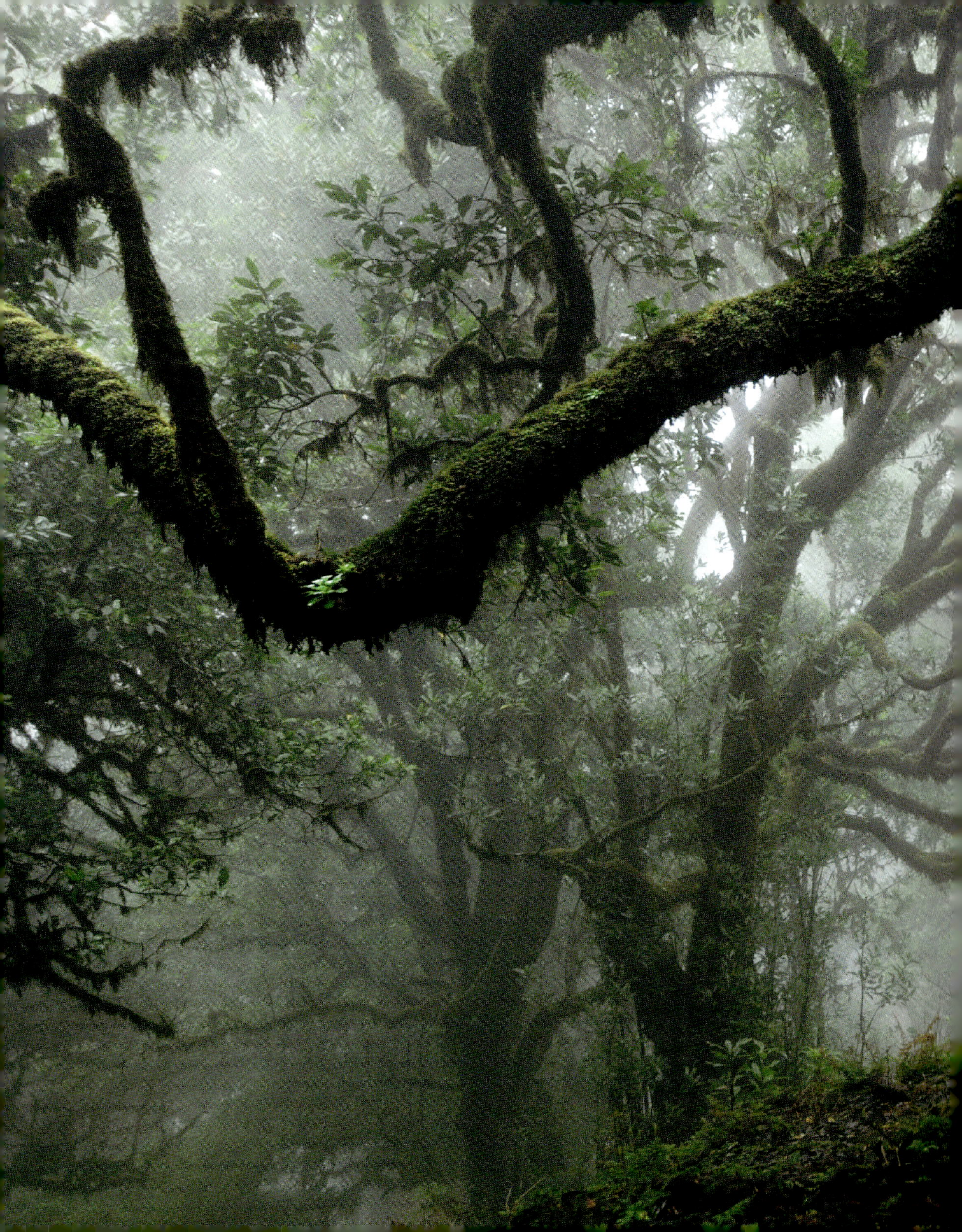

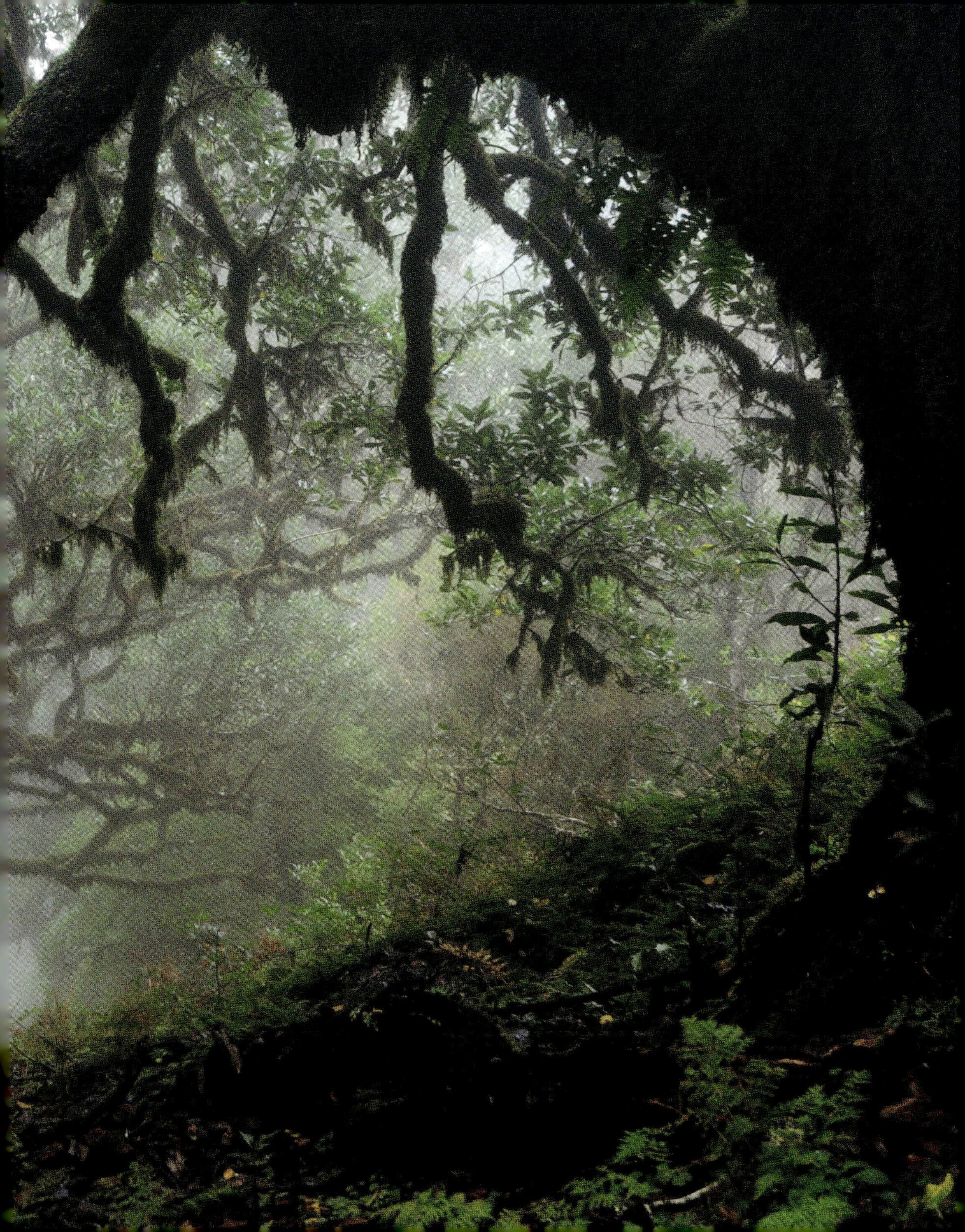

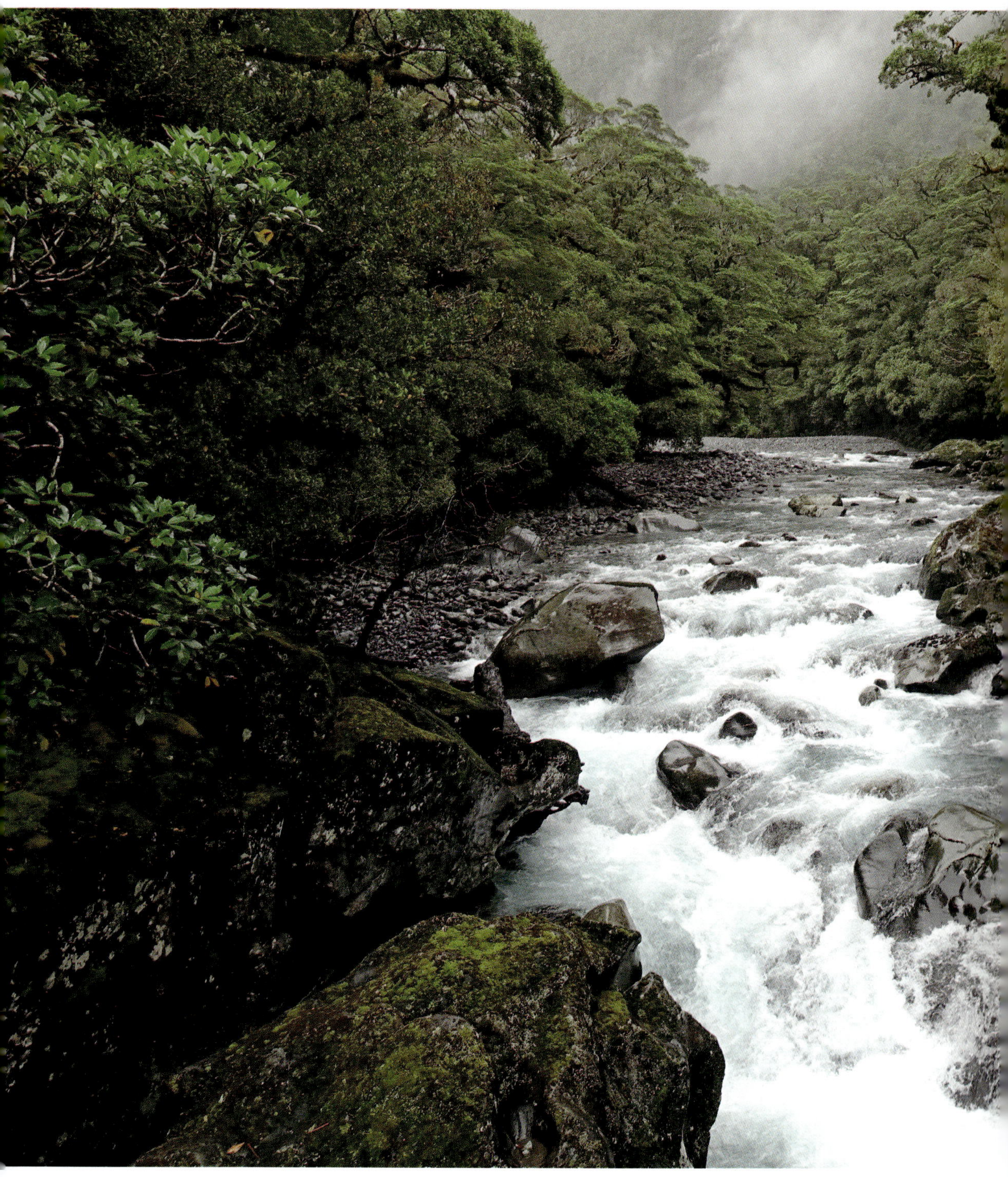

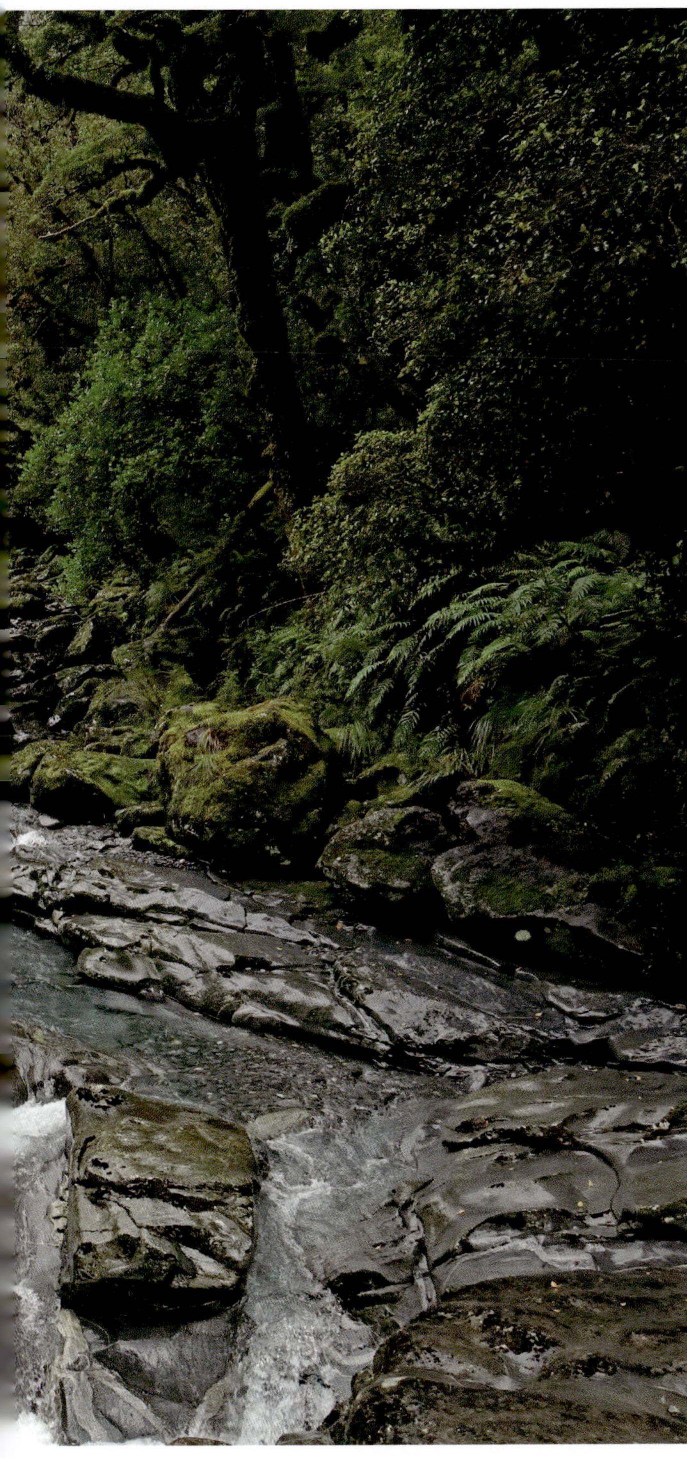

Old trees are vital keystone structures that play unique ecological roles. Beech (*Nothofagus*) forest, New Zealand

Following spread: Violet webcap (*Cortinarius violaceus*), USA

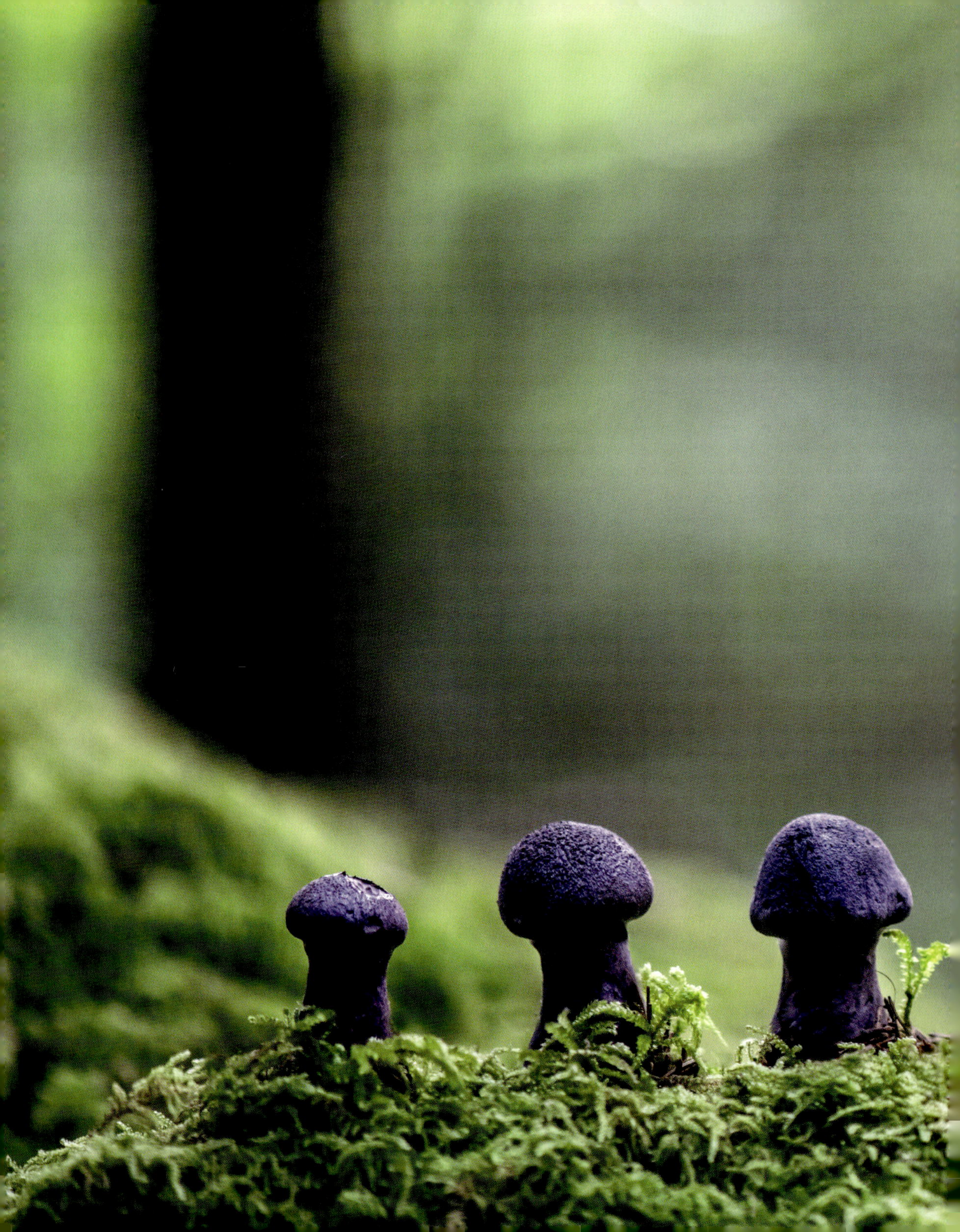

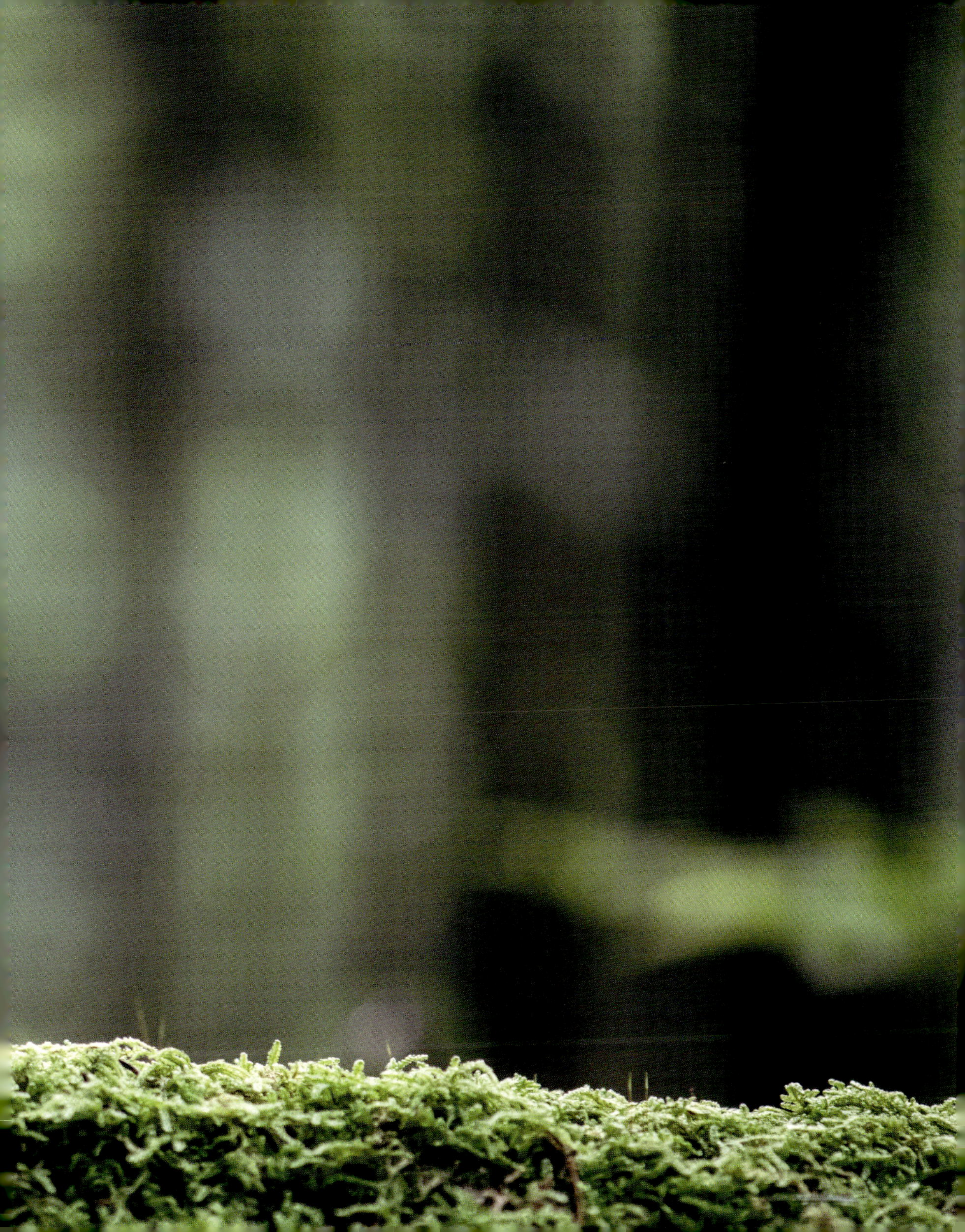

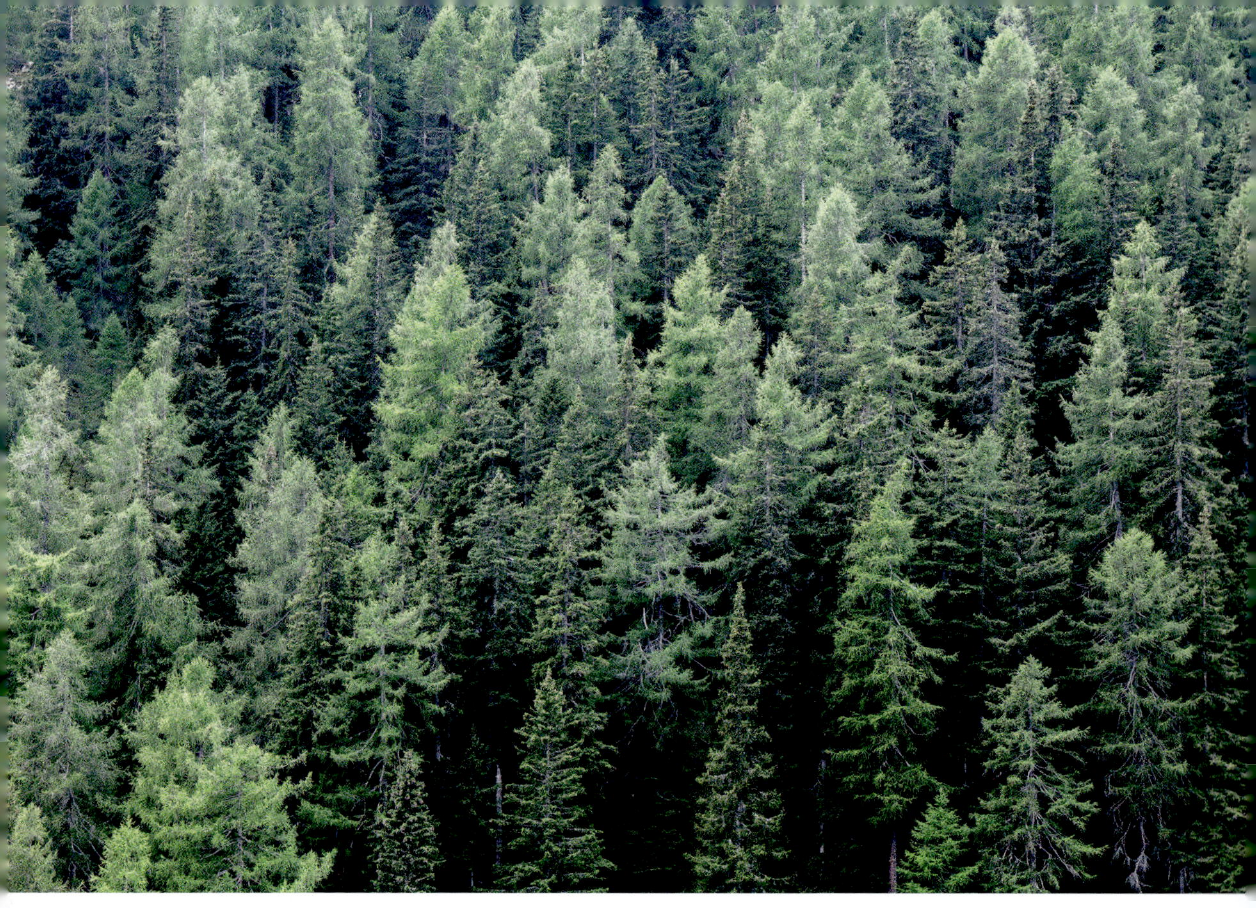

Mixed conifer forest,
Switzerland

Young parasol
(*Macrolepiota procera*),
Switzerland

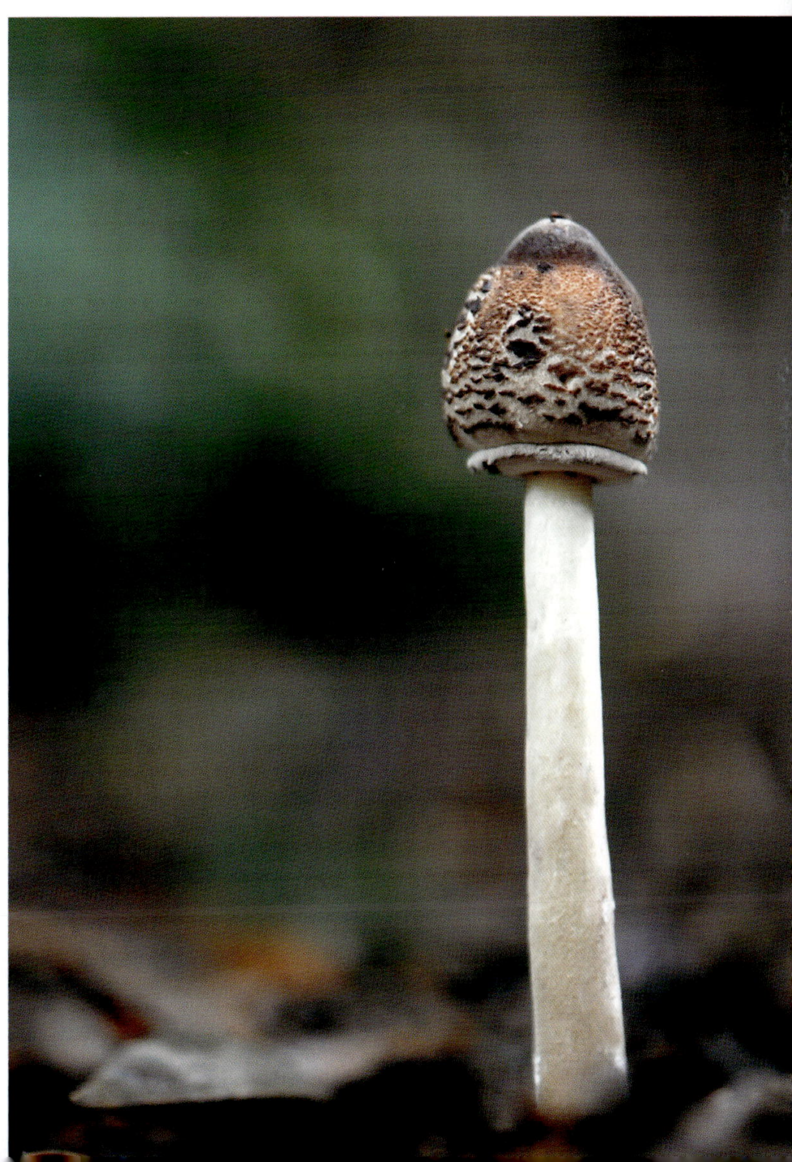

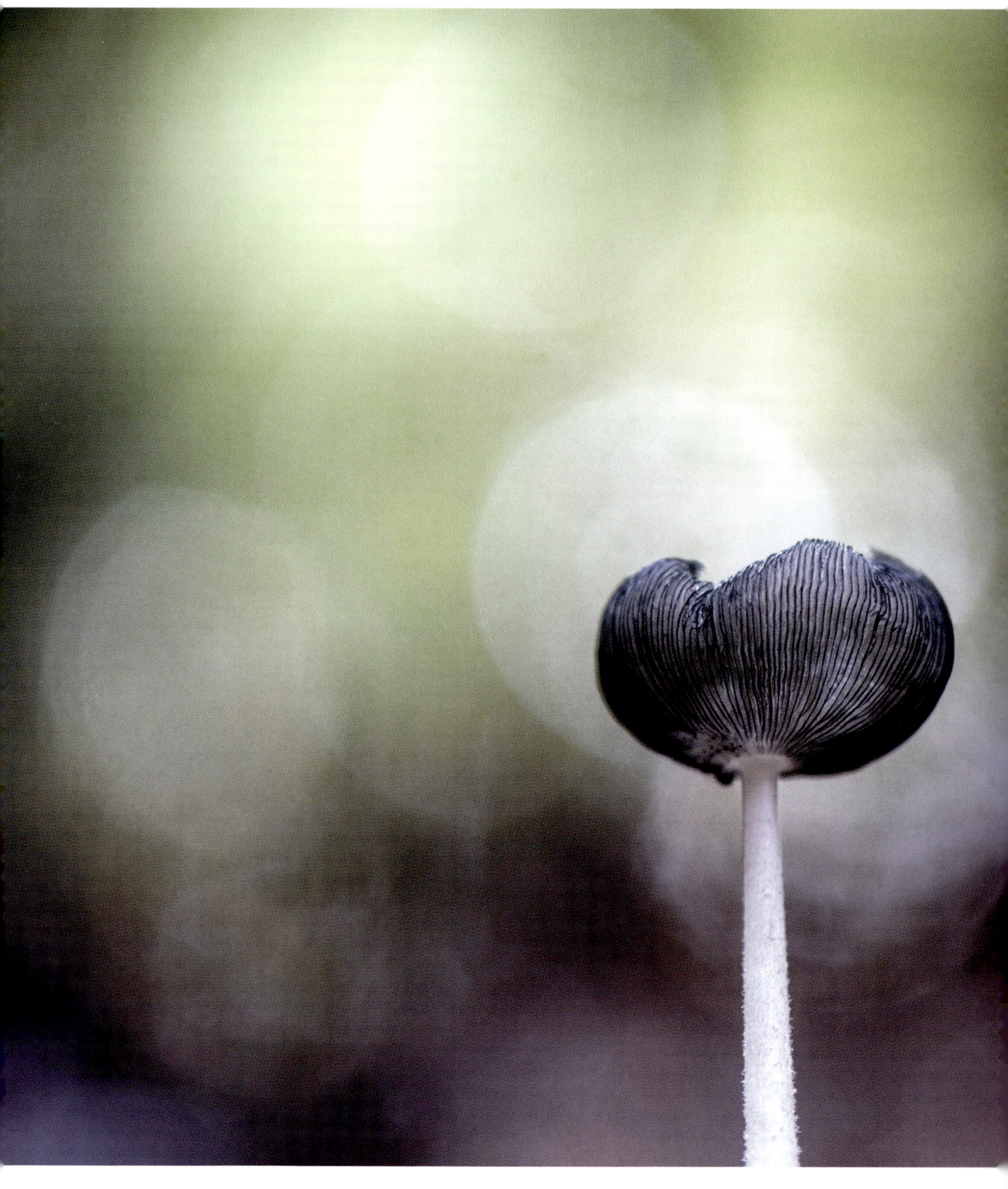

Hare's foot inkcap (*Coprinopsis lagopus*), Canada

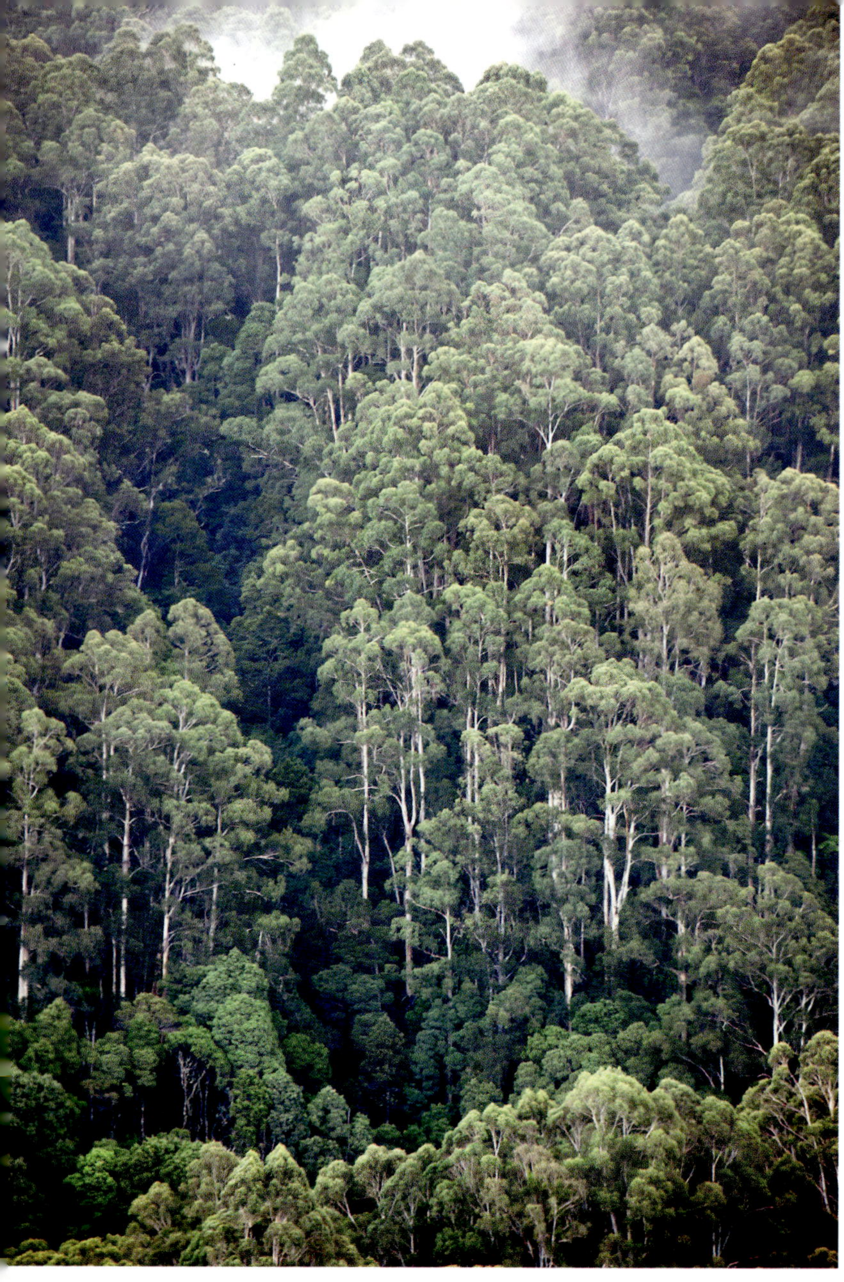

Temperate rainforest, Australia (above)

Stinkwood (*Ocotea foetens*), Madeira, Portugal (opposite)

Following spread: Arolla pine (*Pinus cembra*), Graubünden, Switzerland

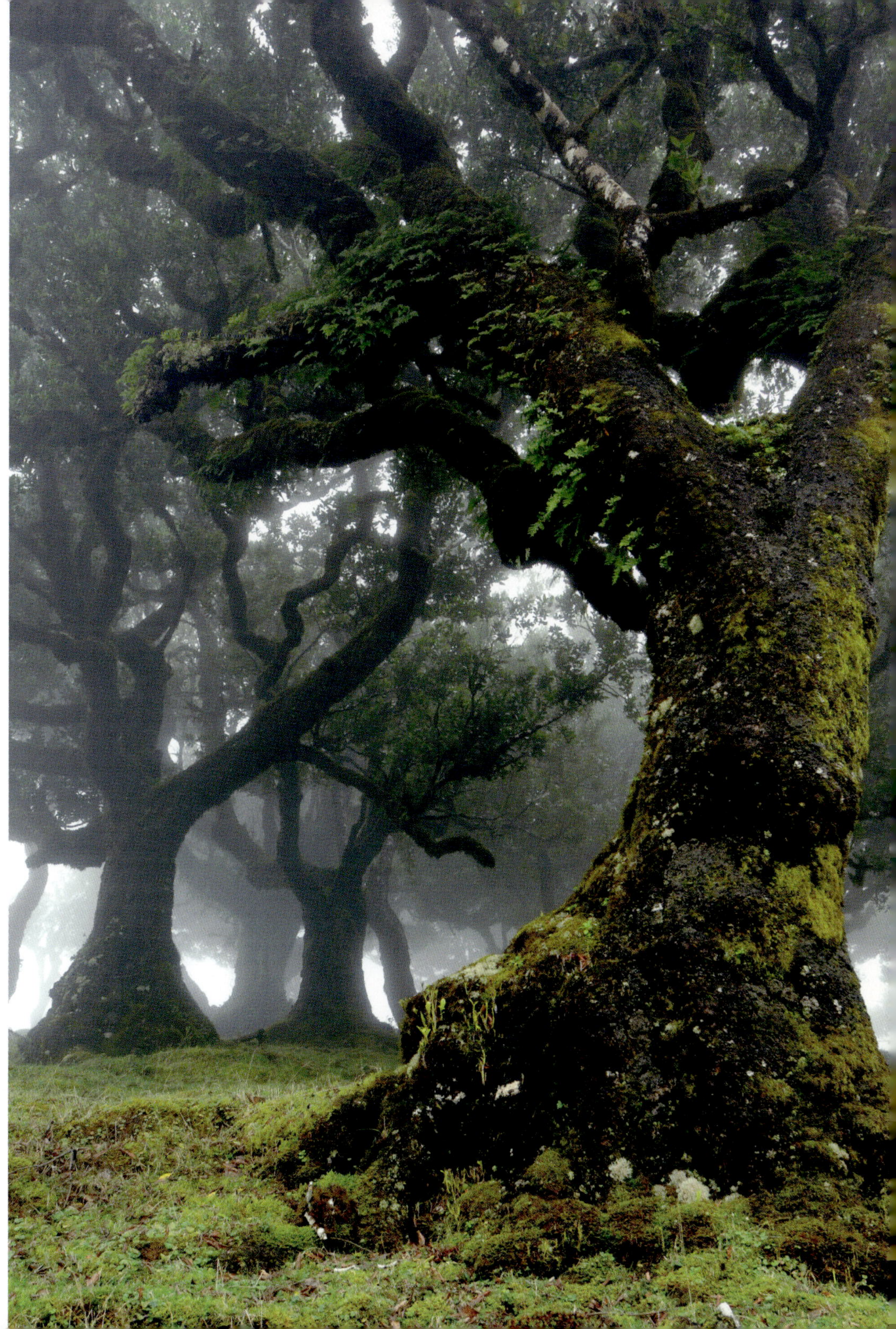

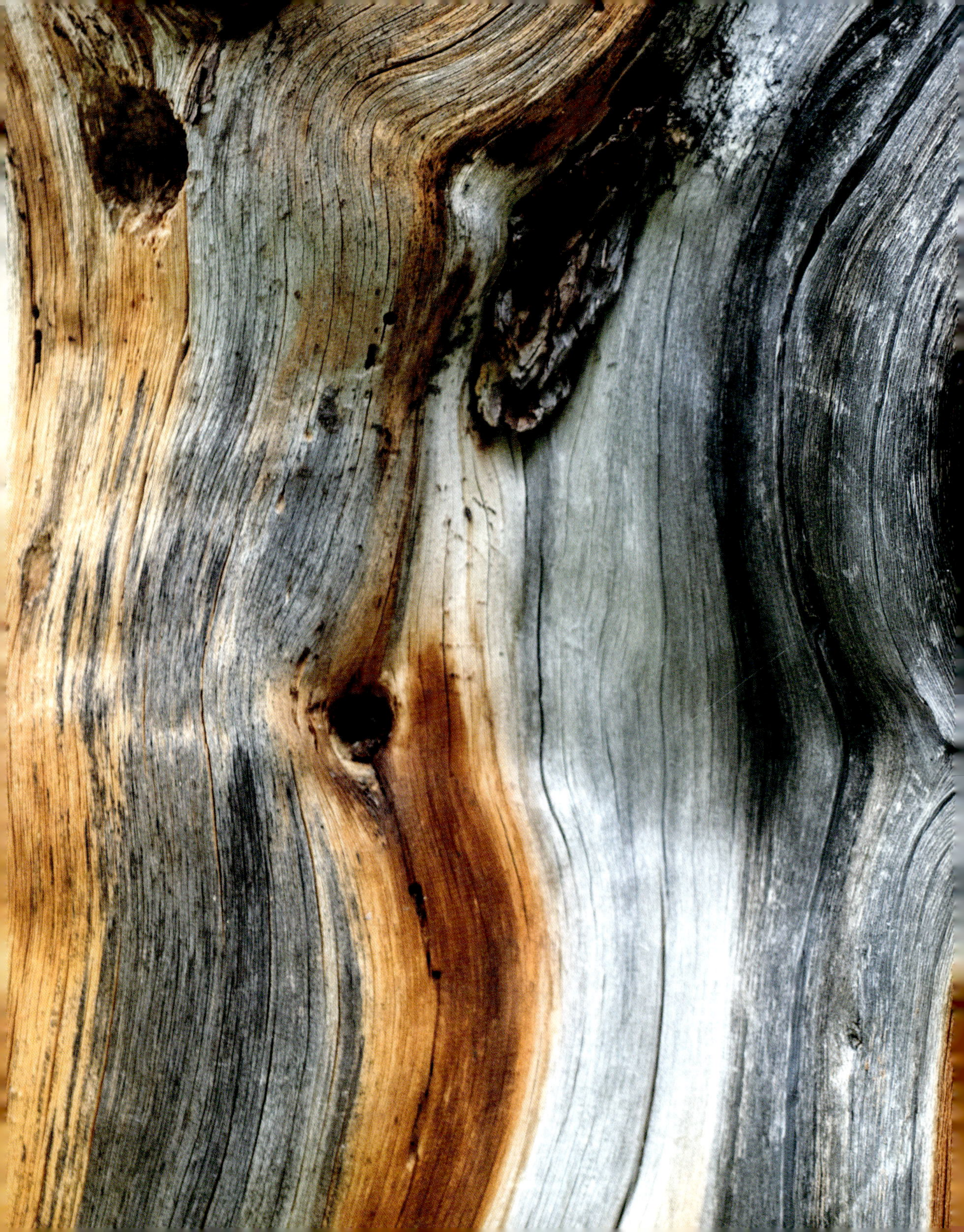

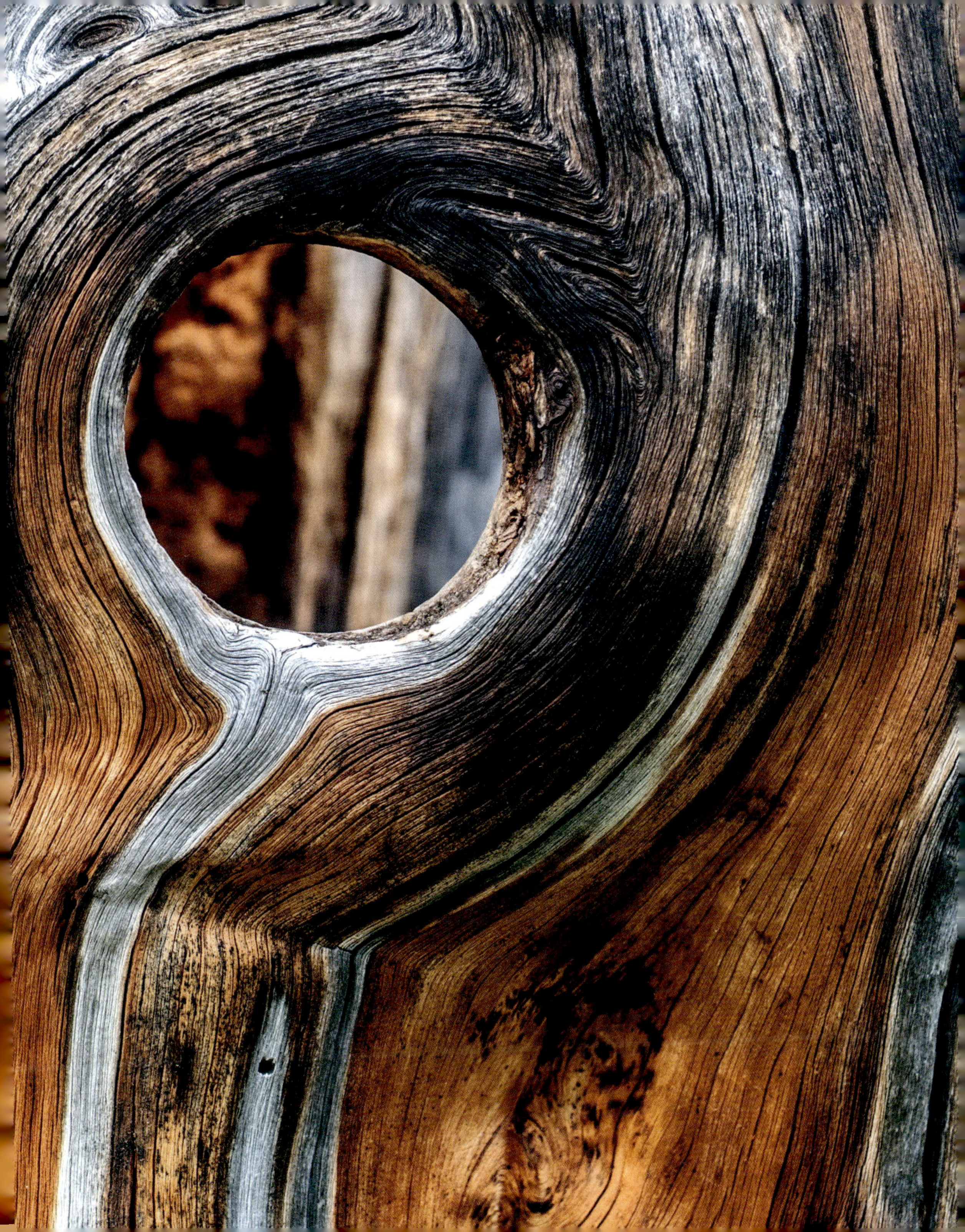

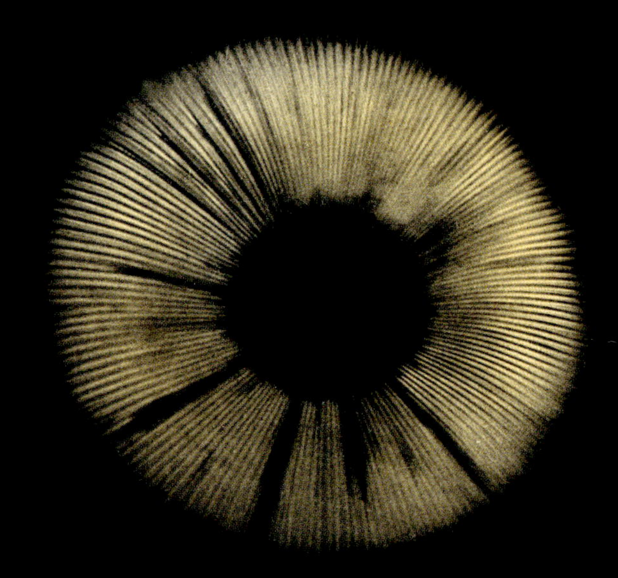

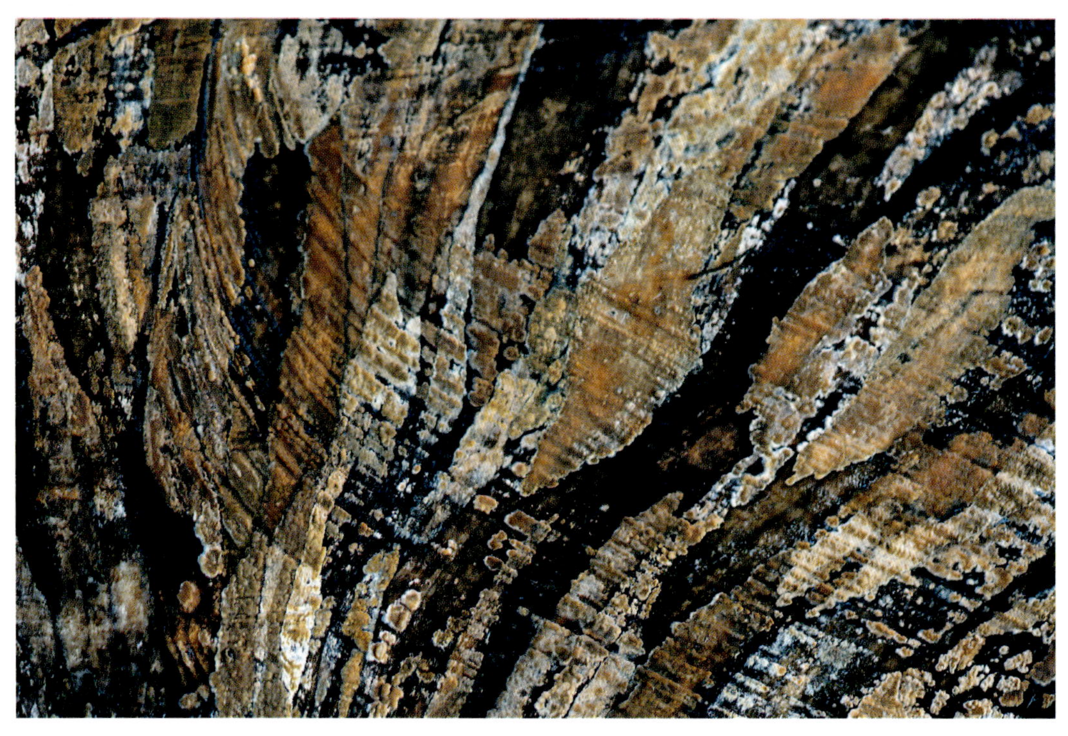

Communities of resupinate fungi, Switzerland

Following spread: Silverleaf fungus (*Chondrostereum purpureum*), USA

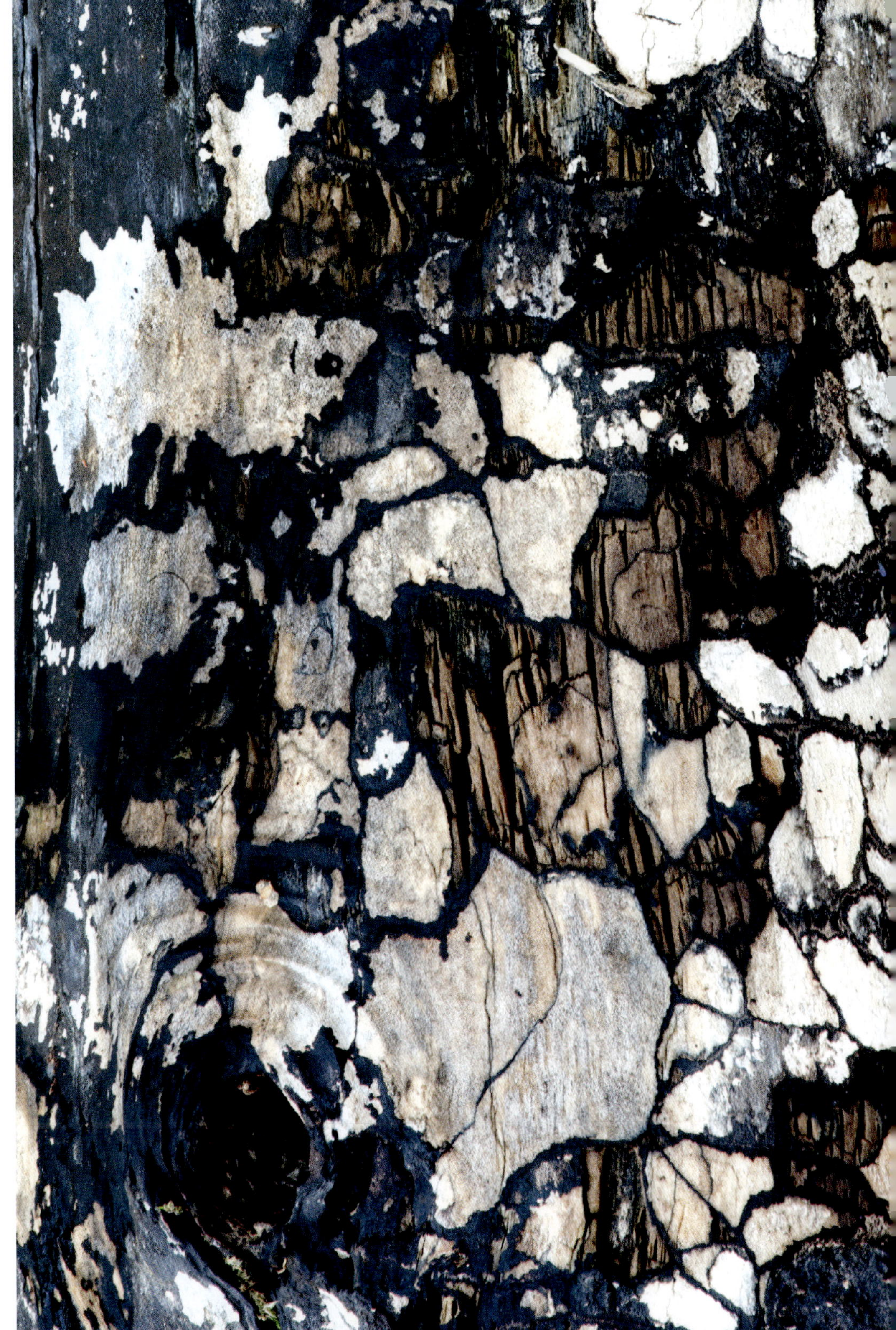

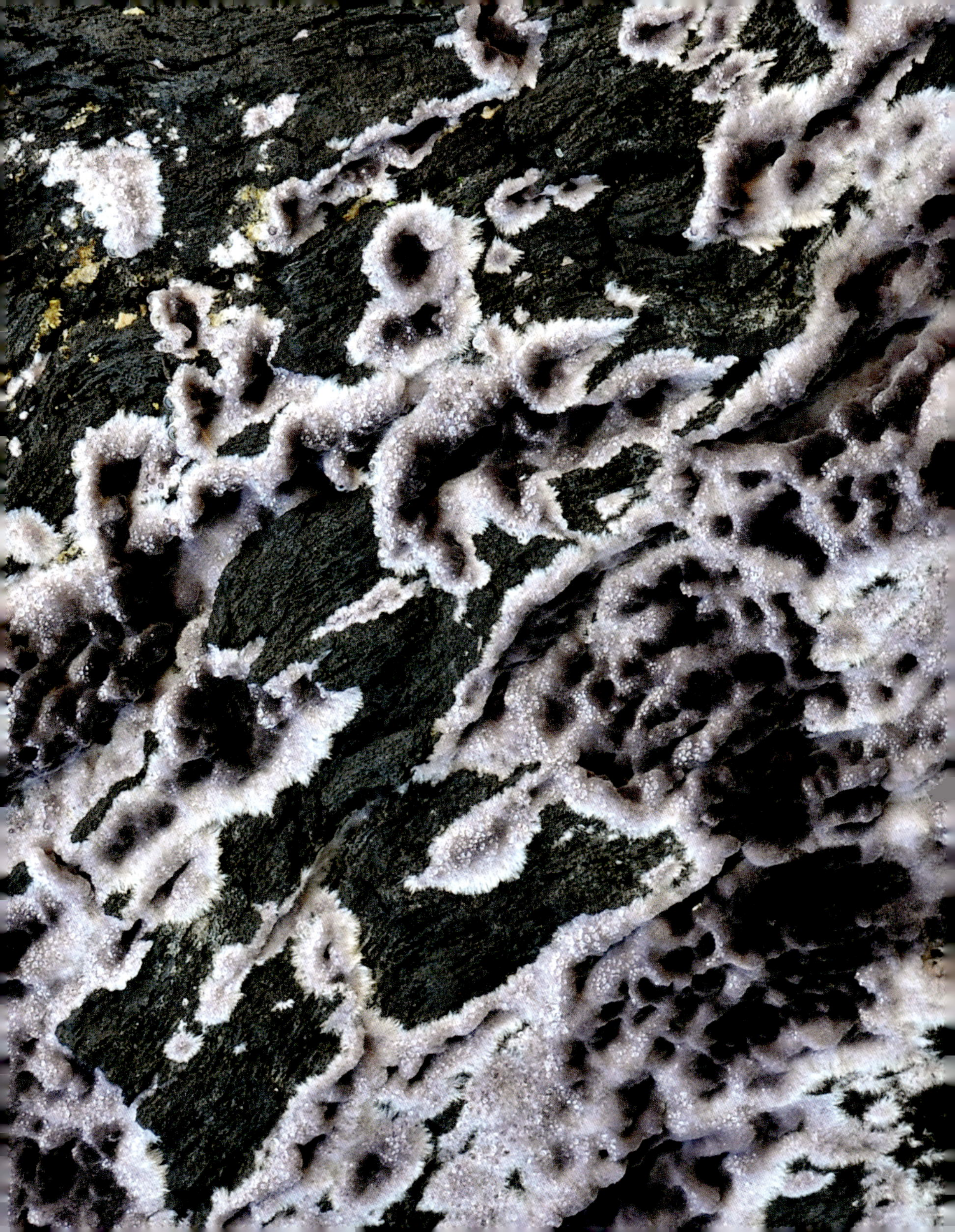

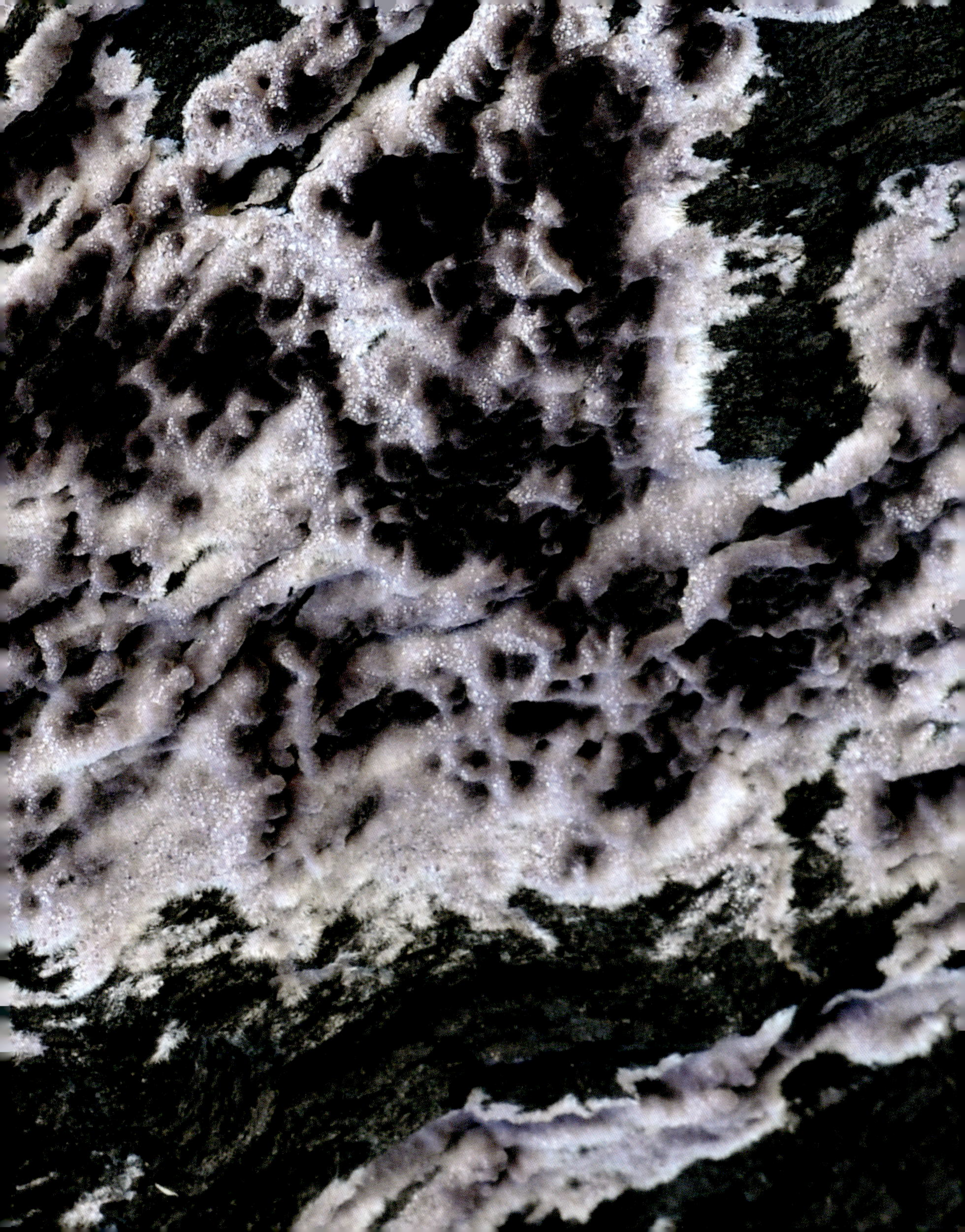

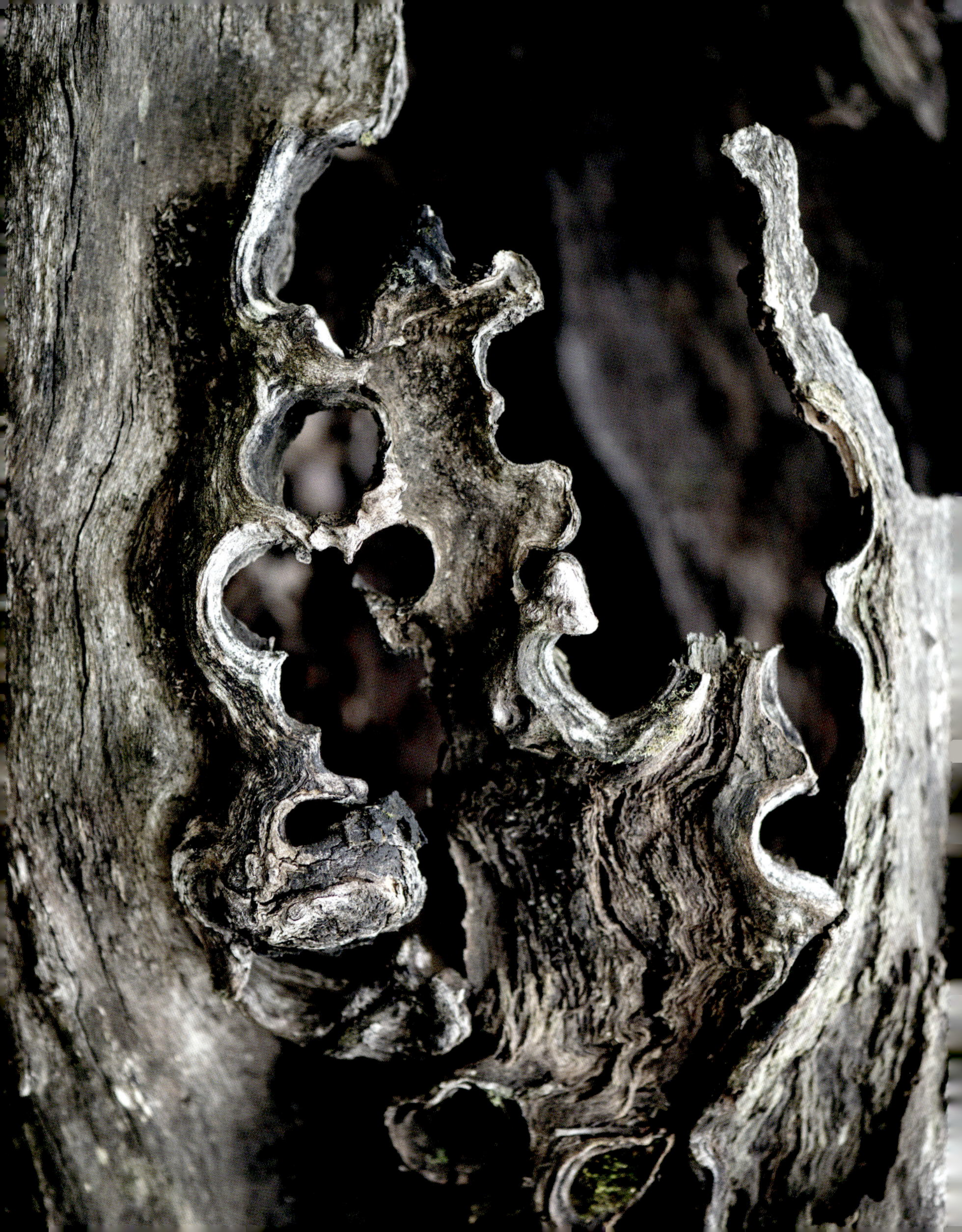

(de)composition

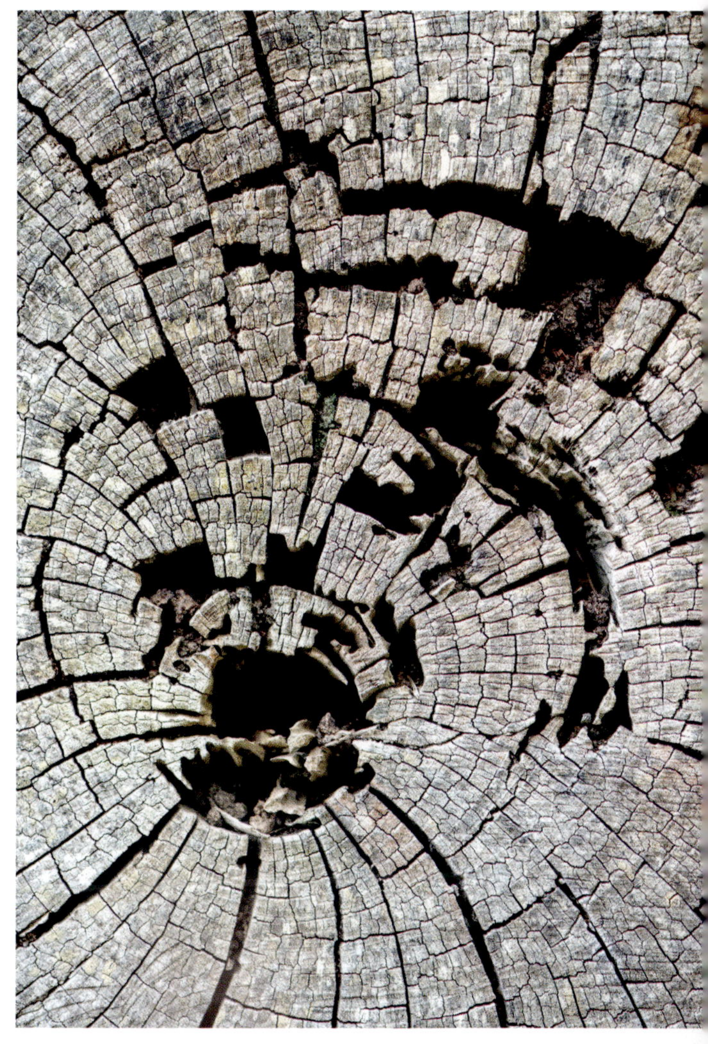

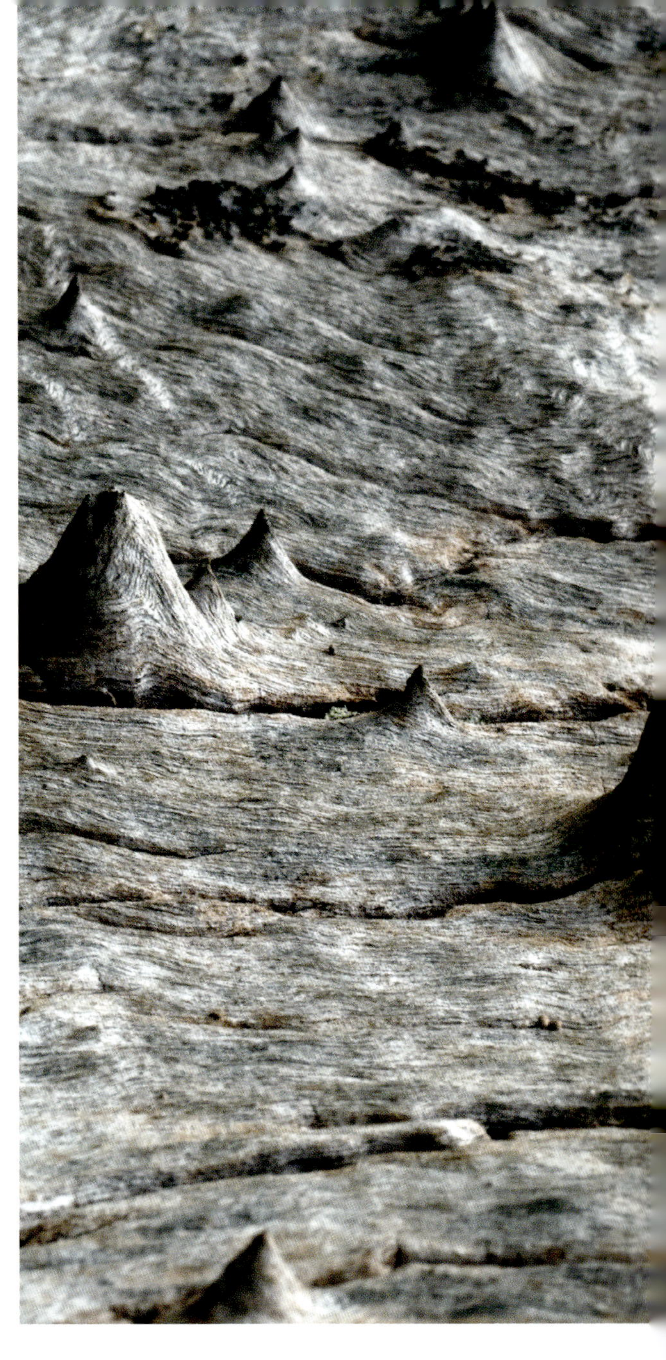

MICROTOPOGRAPHIES

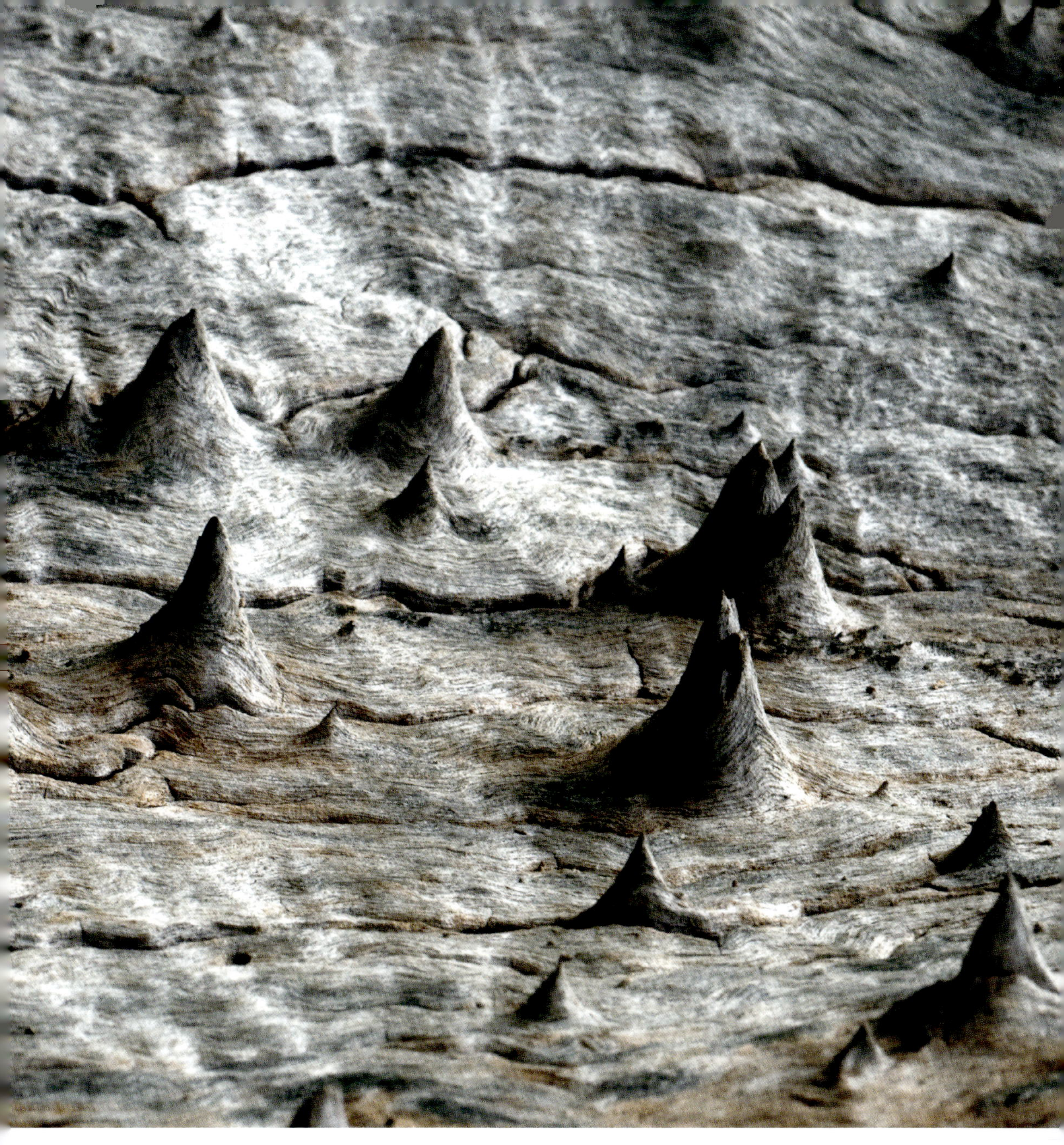

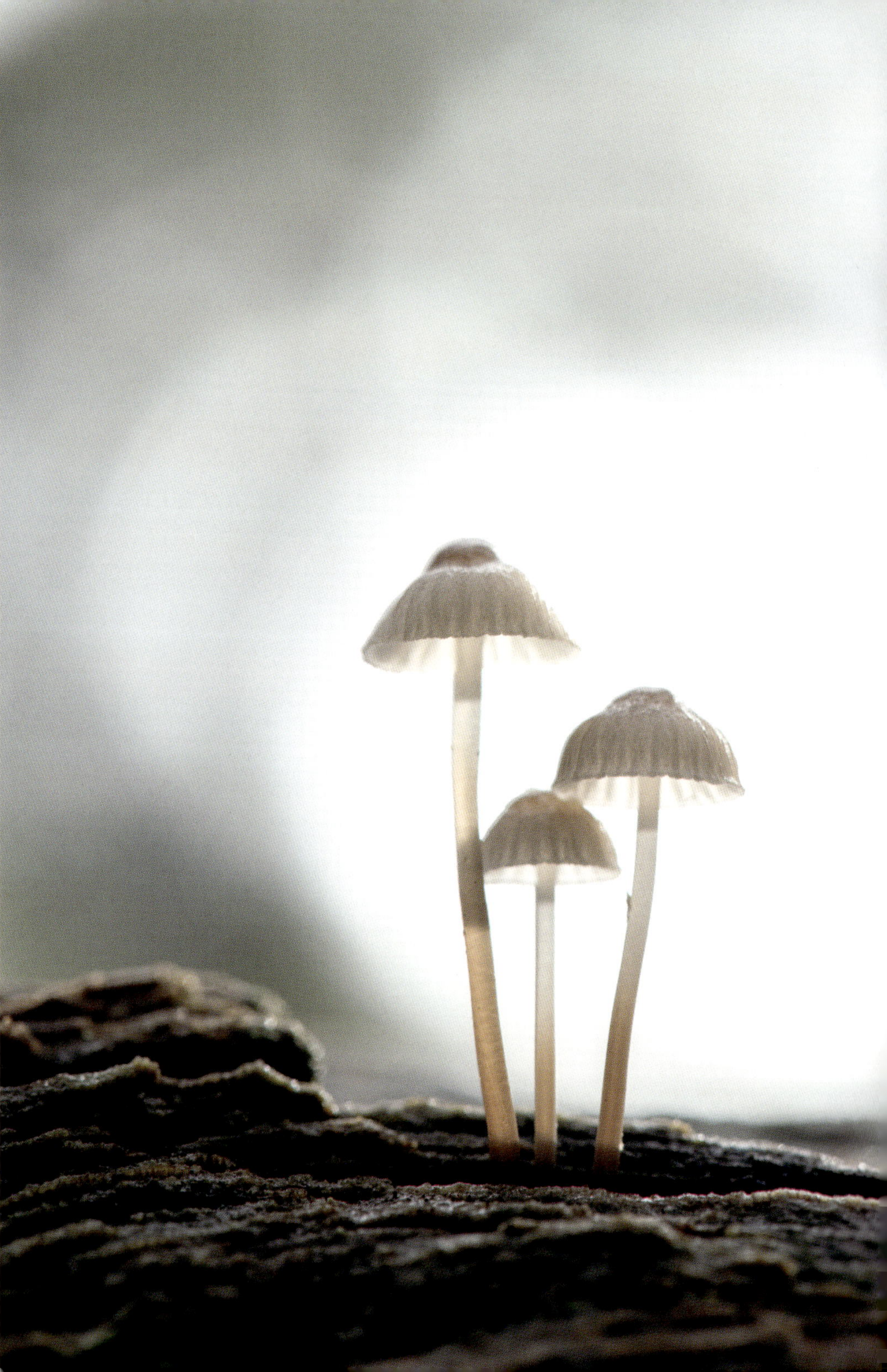

CREVICE SEEKERS

Bonnet mushroom trio
(*Mycena subgalericulata*),
Australia

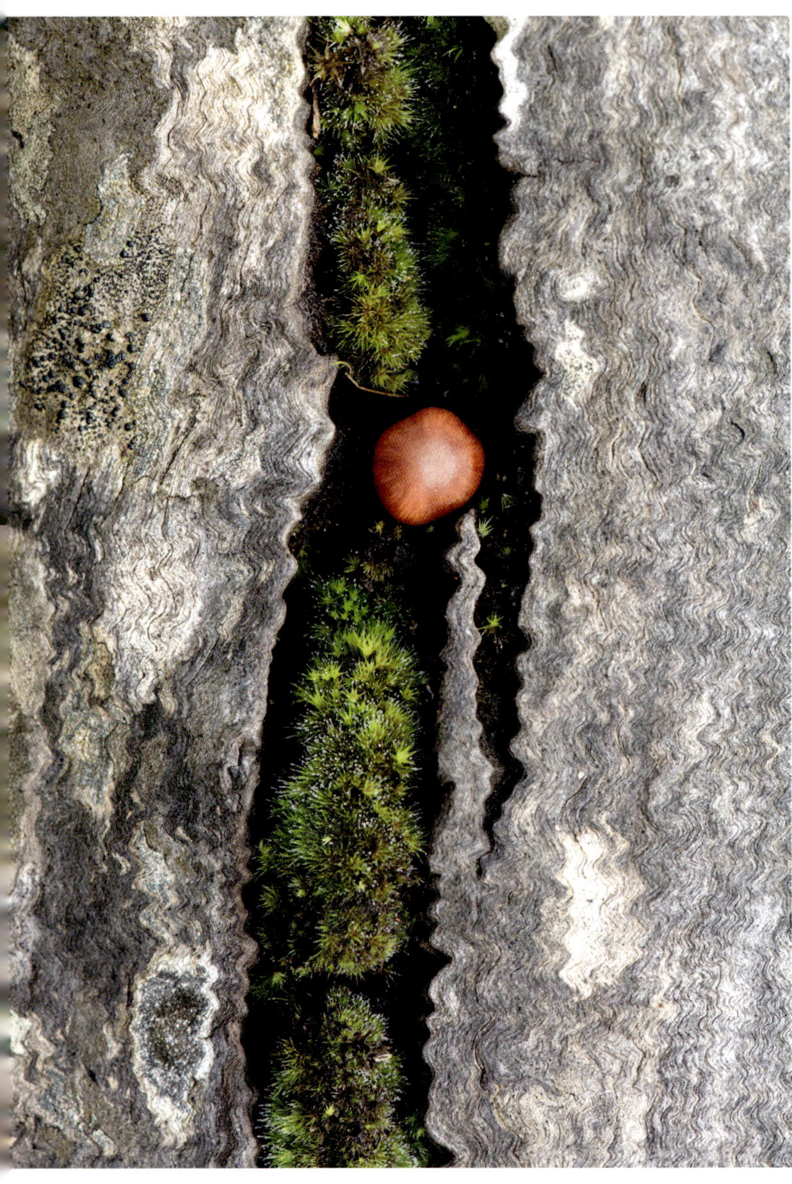

Crevice coloniser (above)

Beetle (*Ogmograptis* sp.) larva tracks in scribbly gum (*Eucalyptus rossii*), Australia (opposite)

Following spread: Spalted wood, Poschiavo valley, Switzerland

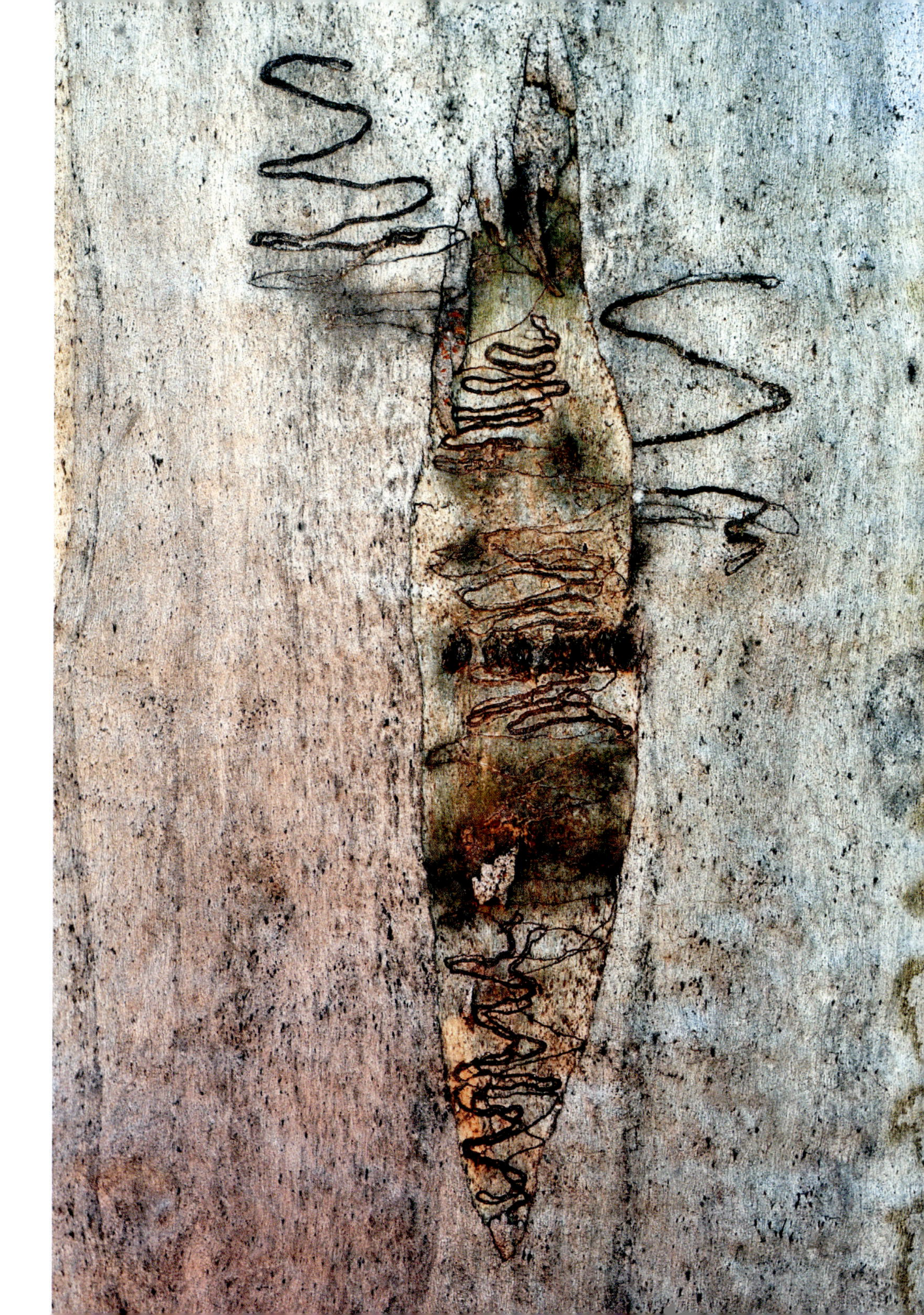

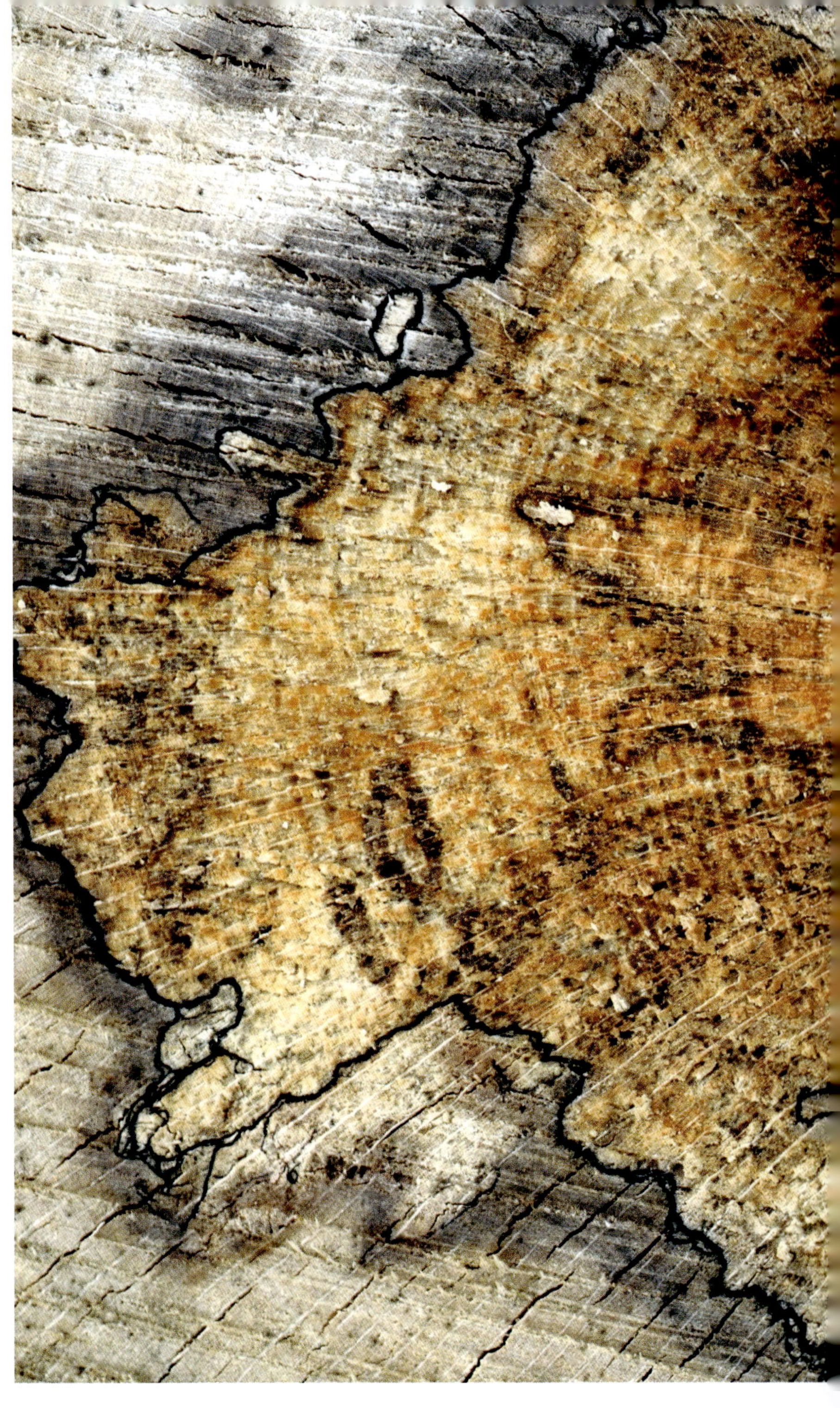

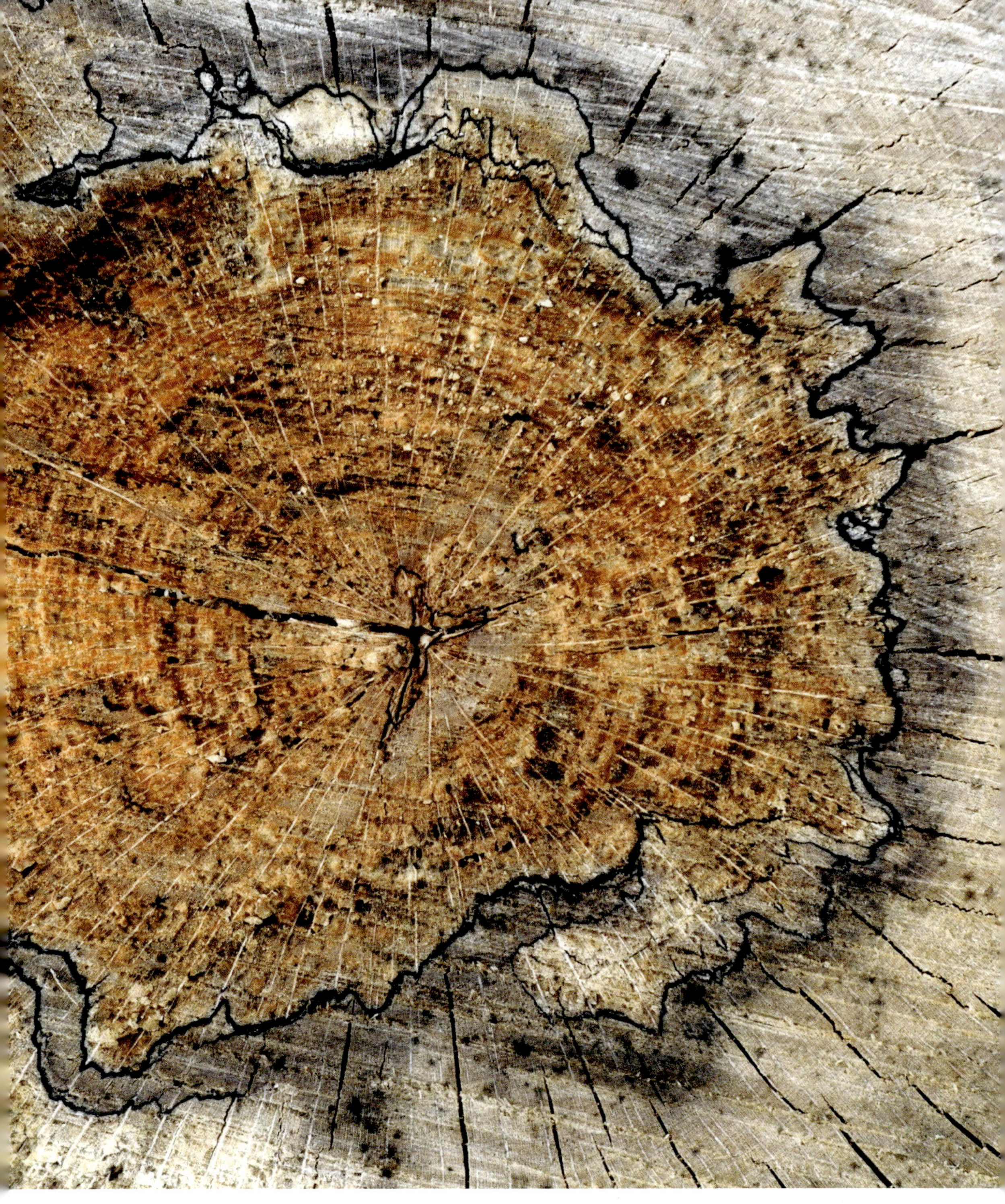

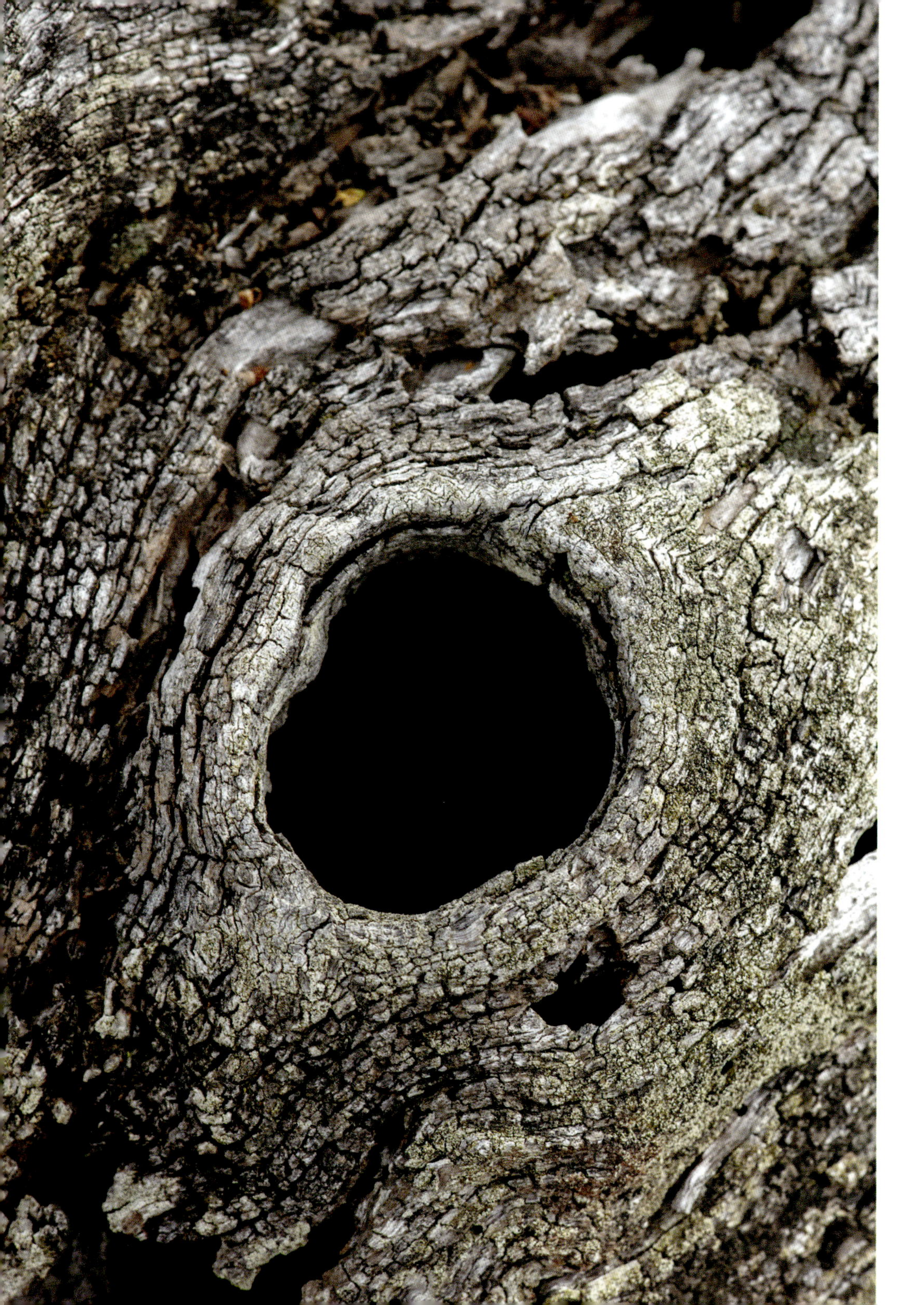

hollow

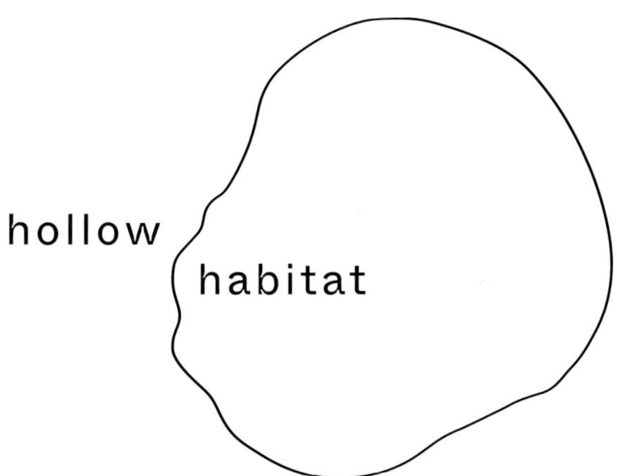

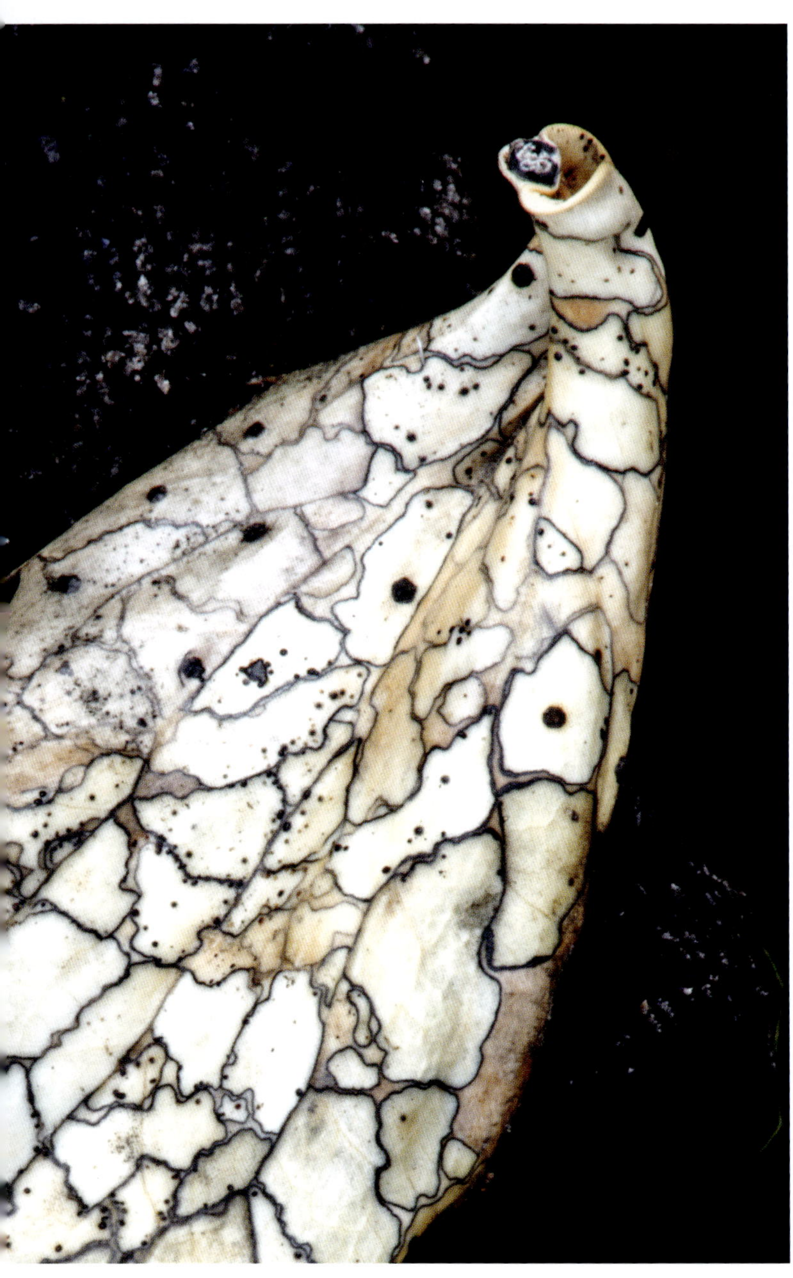

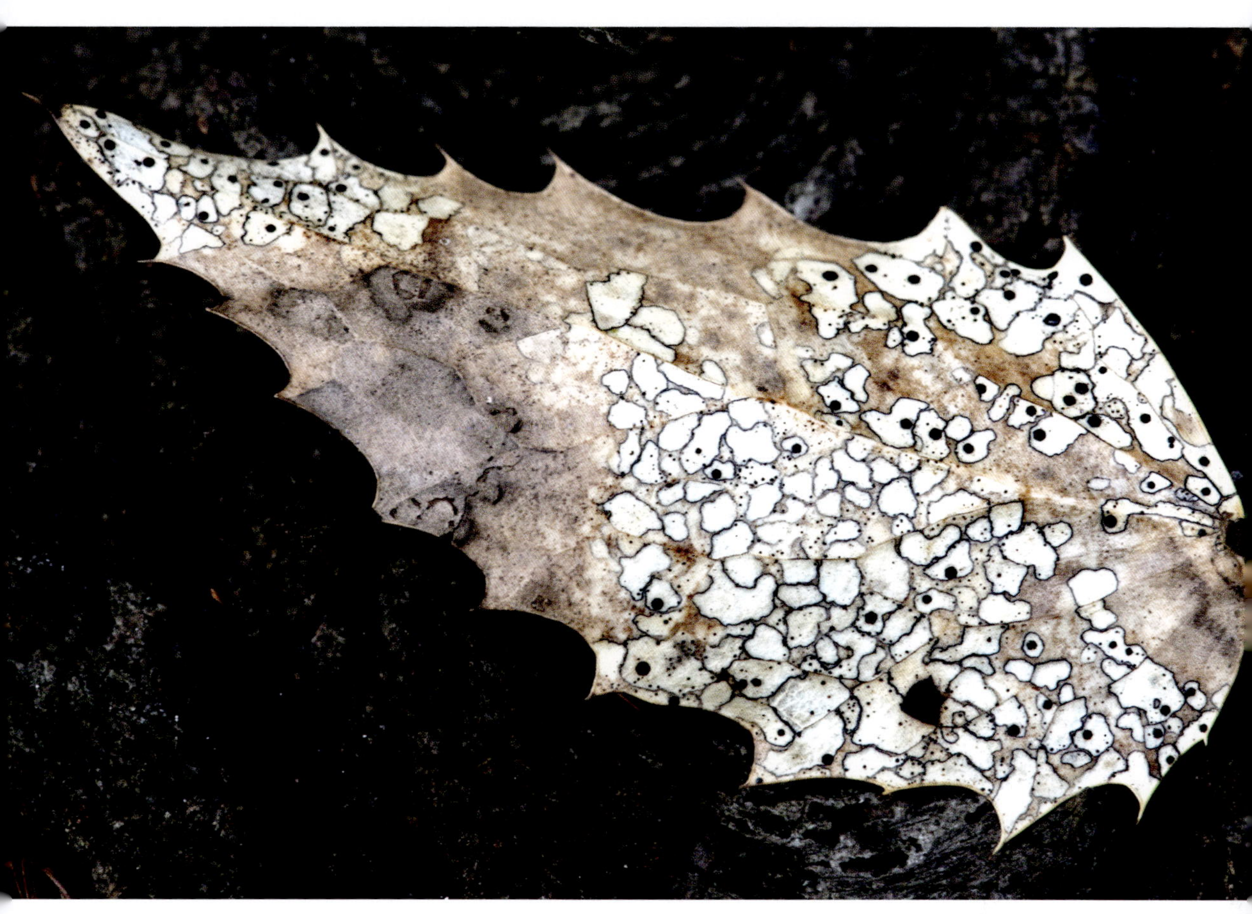

Map fungus
(*Coccomyces dentatus*)
on Oregon grape
(*Berberis aquifolium*)
leaf, USA

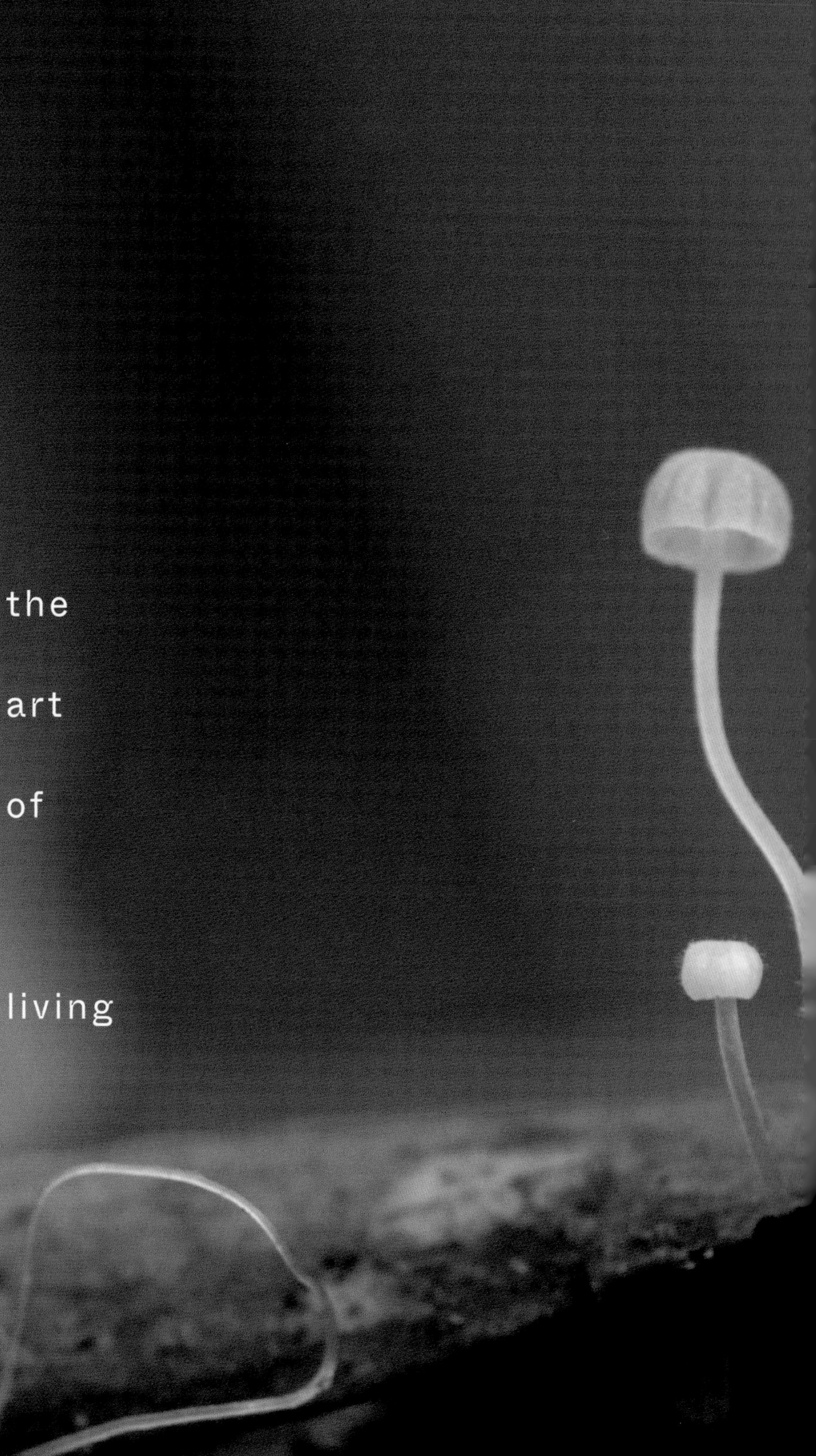

the art of leaf living

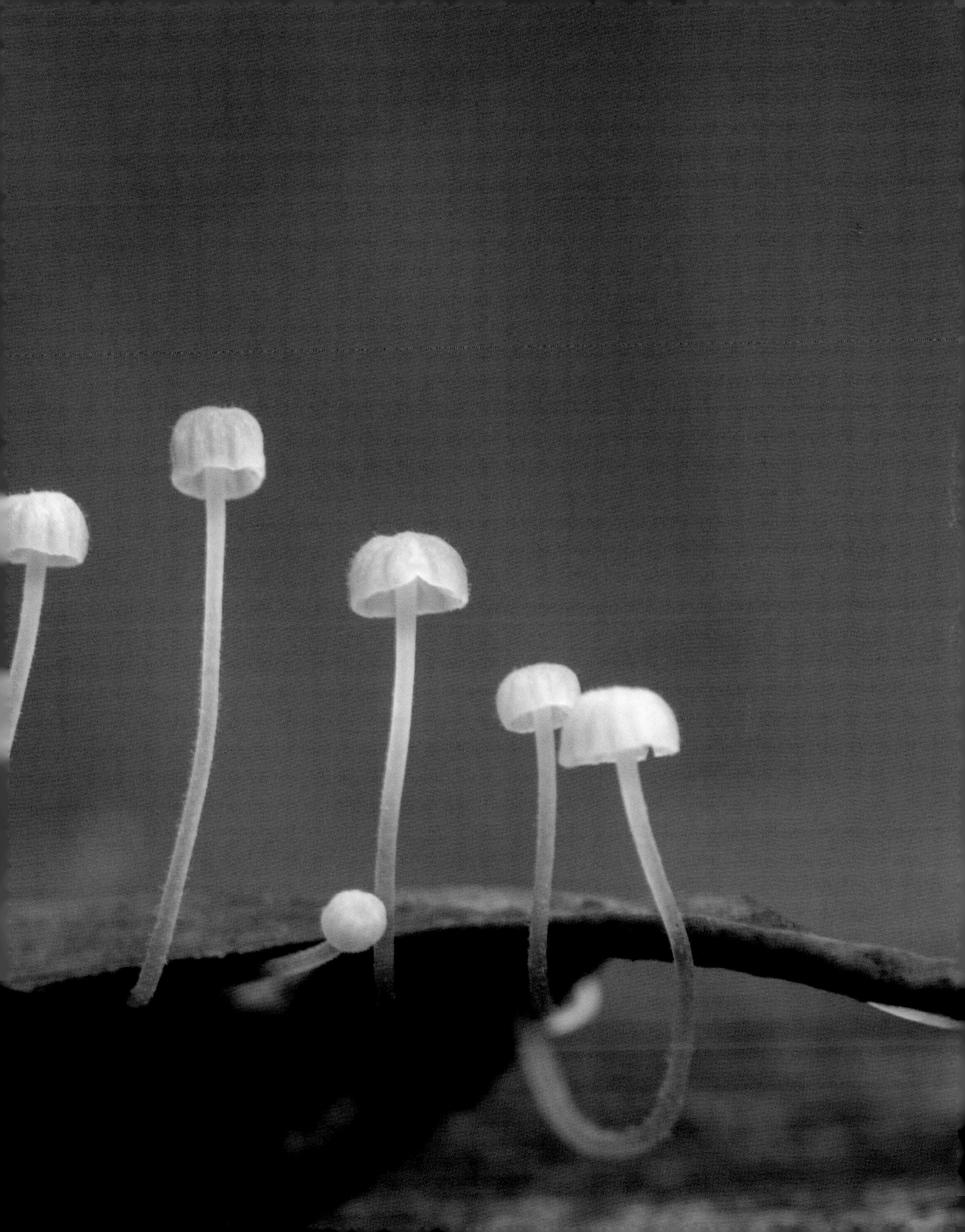

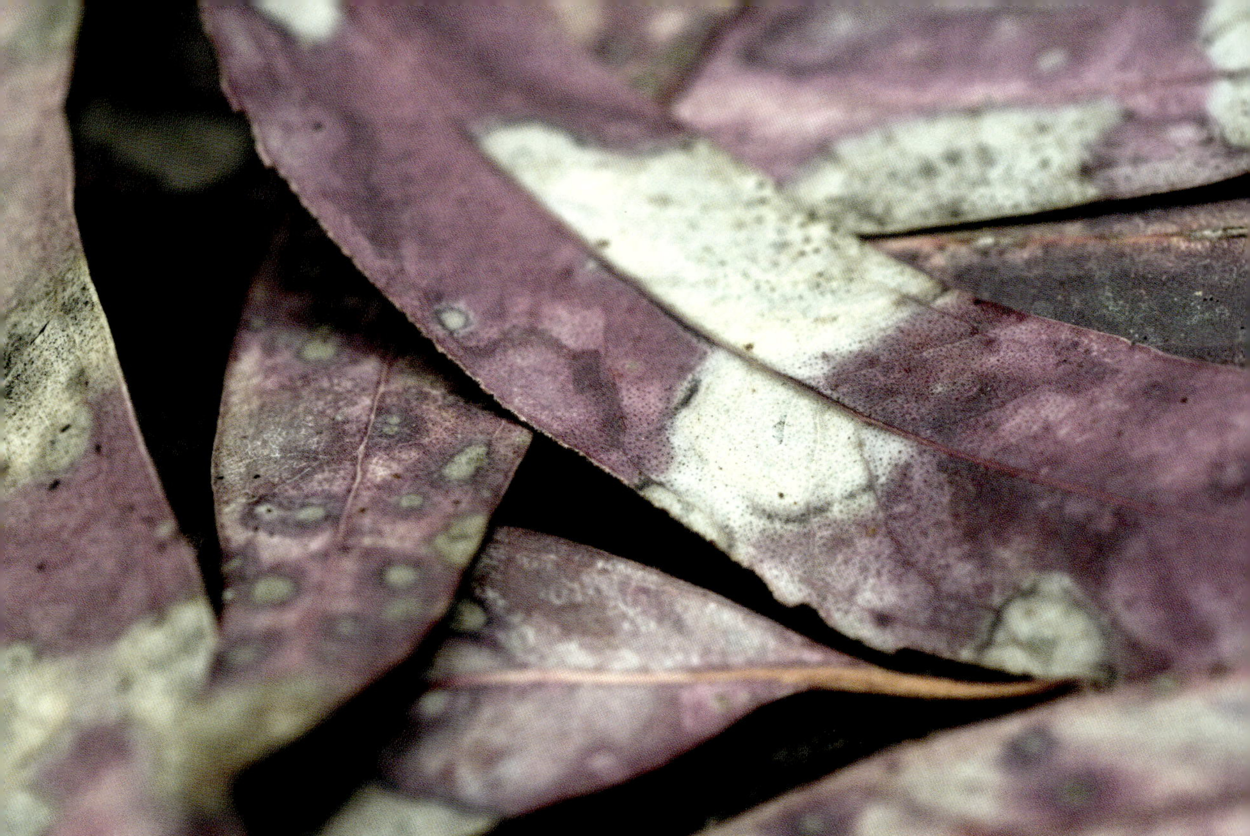

Fungus food, Australia

Bonnet mushroom
(*Mycena* sp.), Switzerland

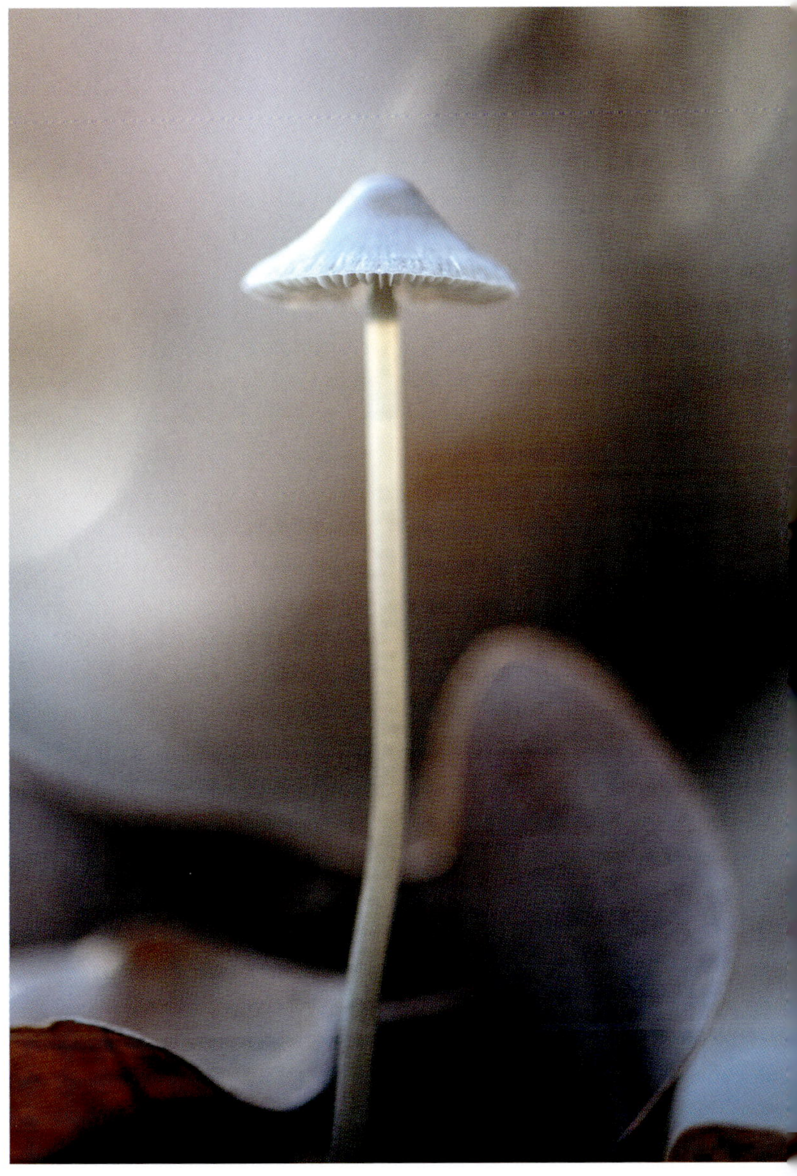

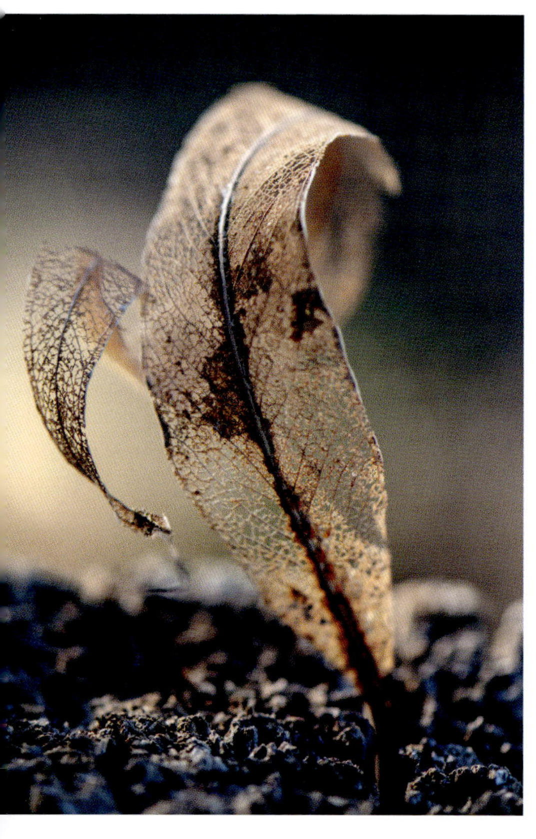

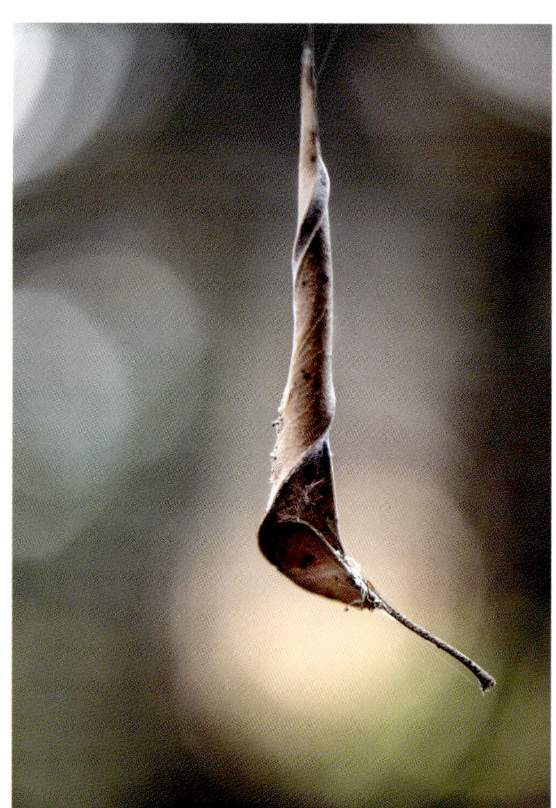

An individual leaf

harbours myriad organisms

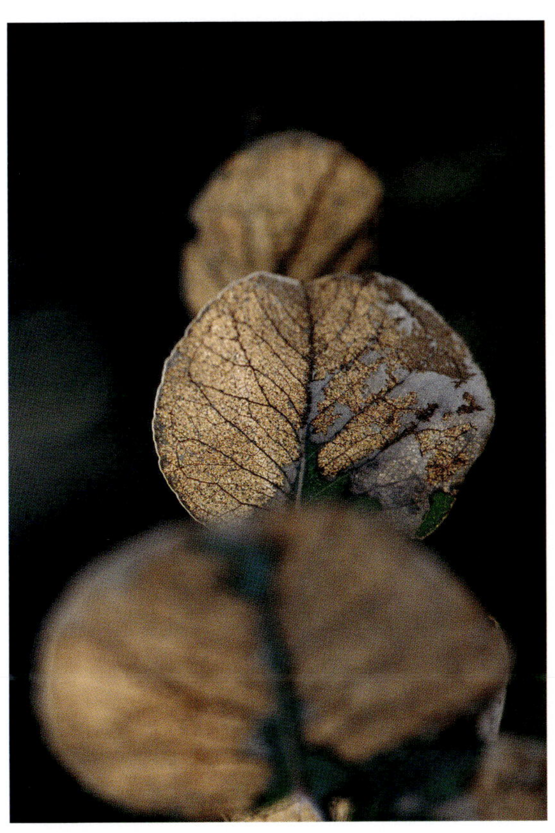

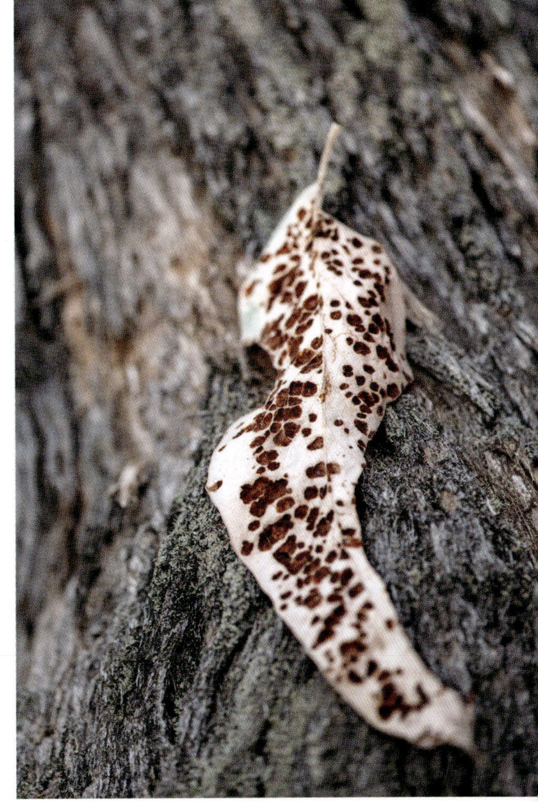

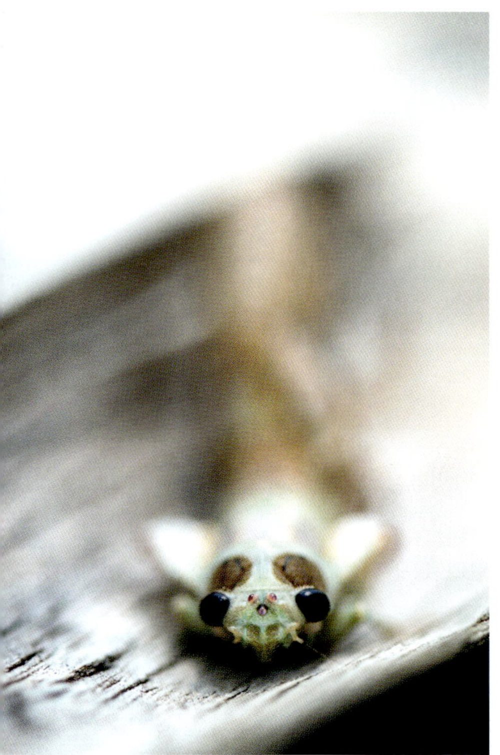

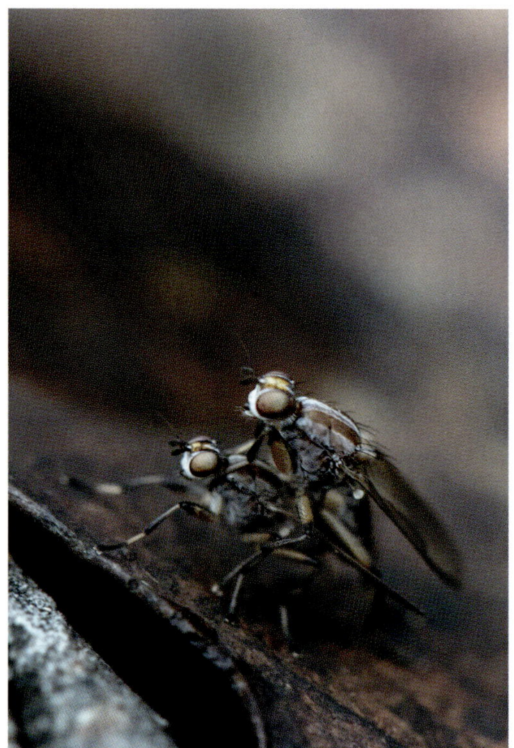

among them the spineless

and the fungal

Left to right:

Emerging green grocer cicada (*Cyclochila australasiae*), Australia

Amorous fungus flies (*Tapeigaster*), Australia

Litter life (opposite)

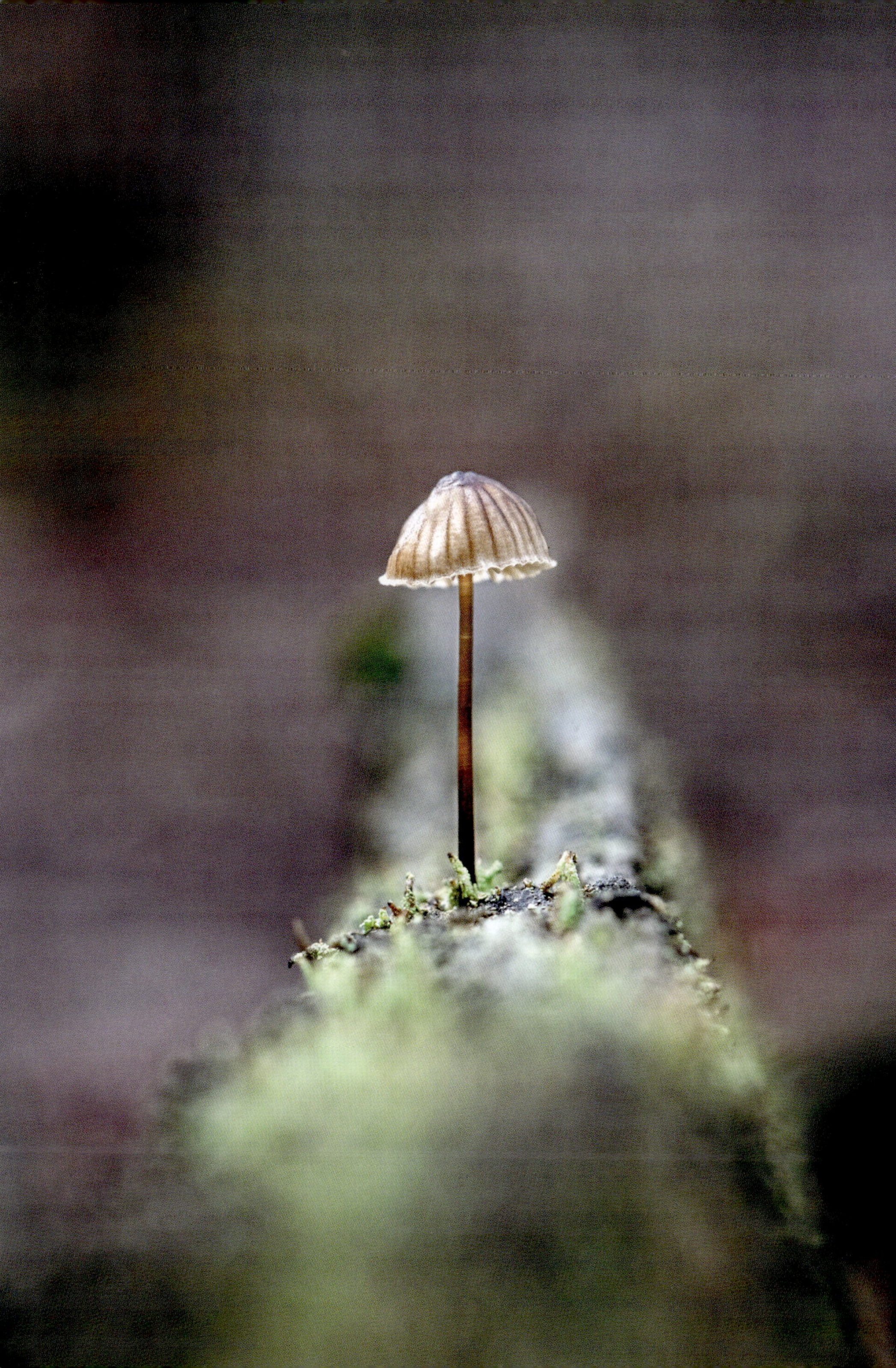

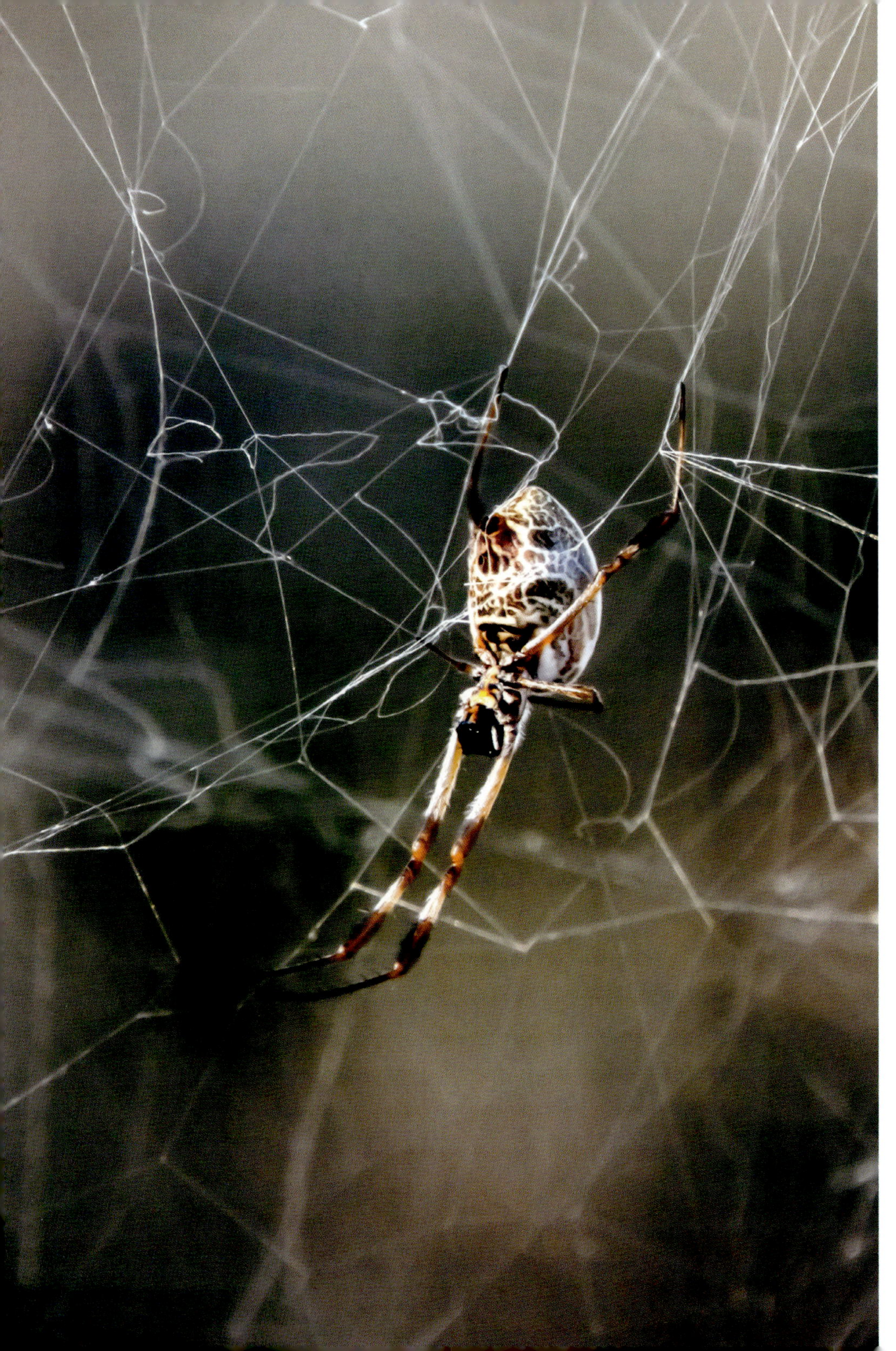

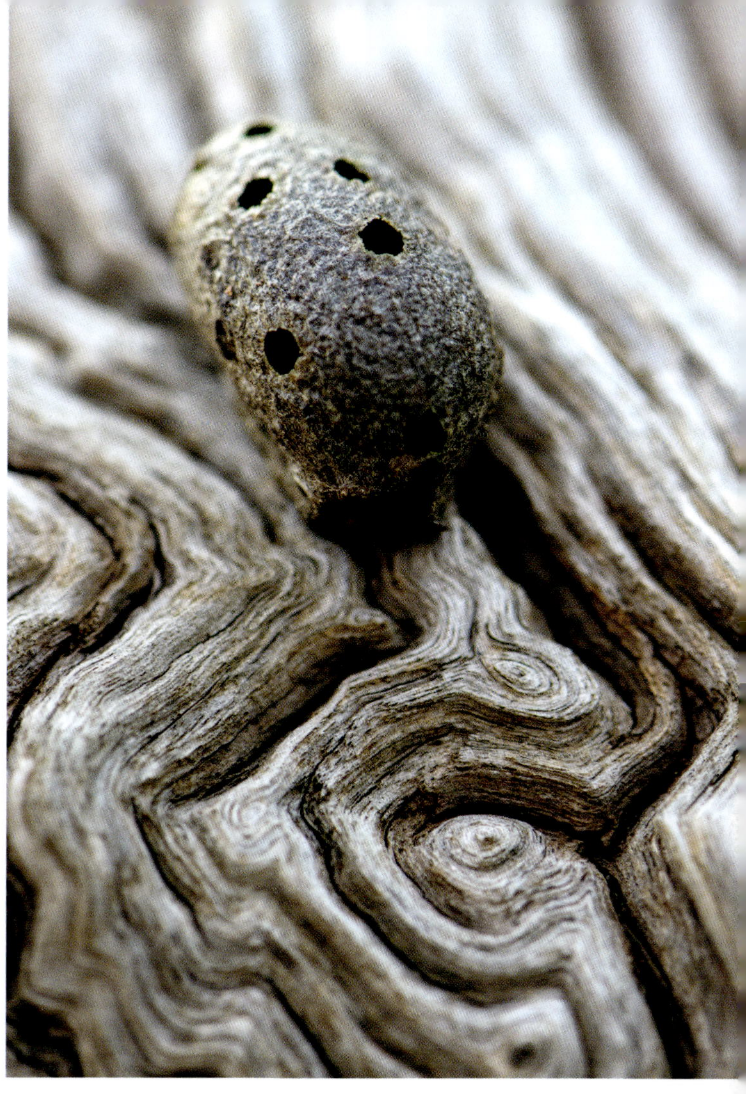

Golden orb weaver (*Trichonephila edulis*), Australia (opposite)

Emperor gum moth (*Opodiphthera eucalypti*) cocoon with wasp holes, Australia (above)

Following spread: Mycelium

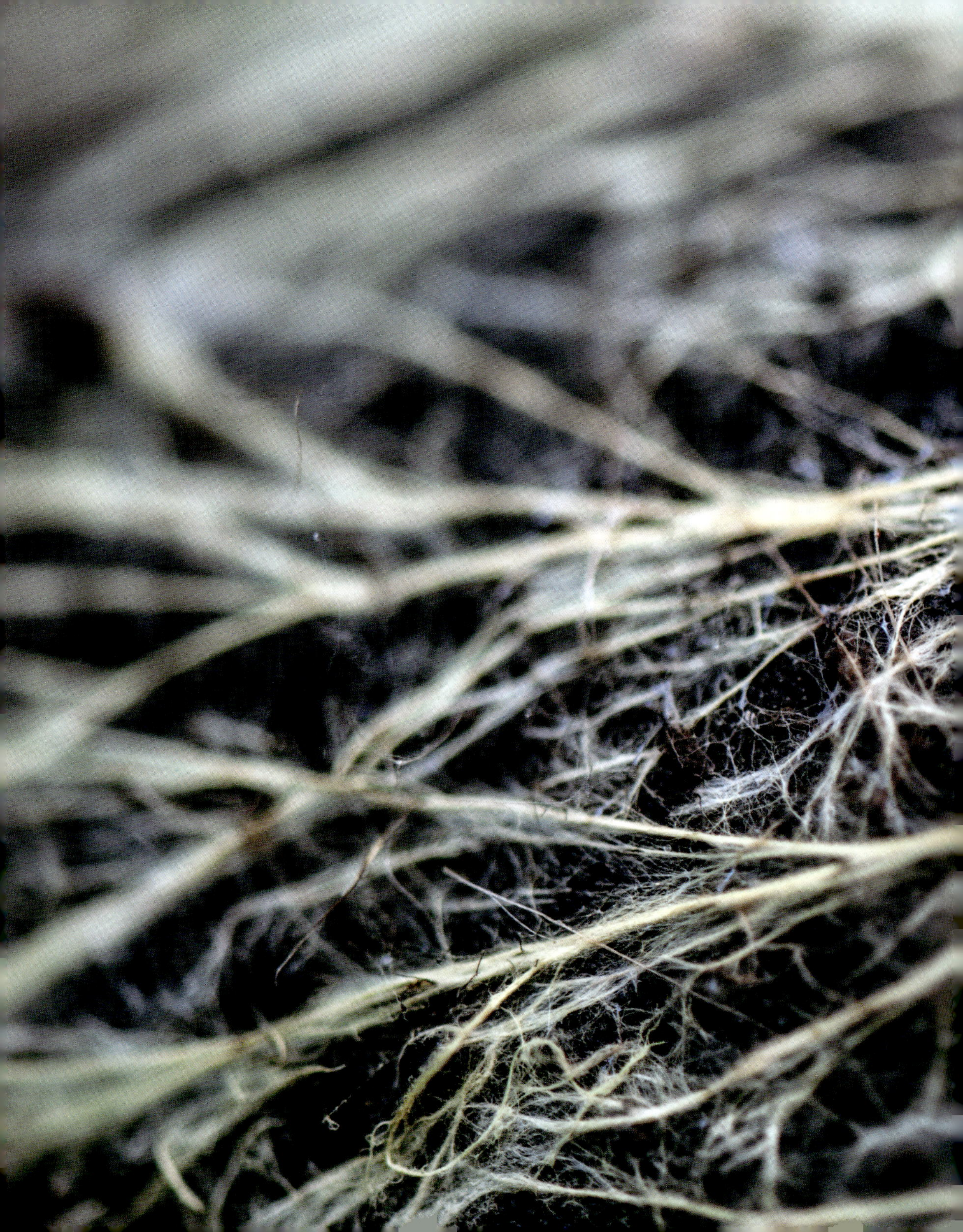

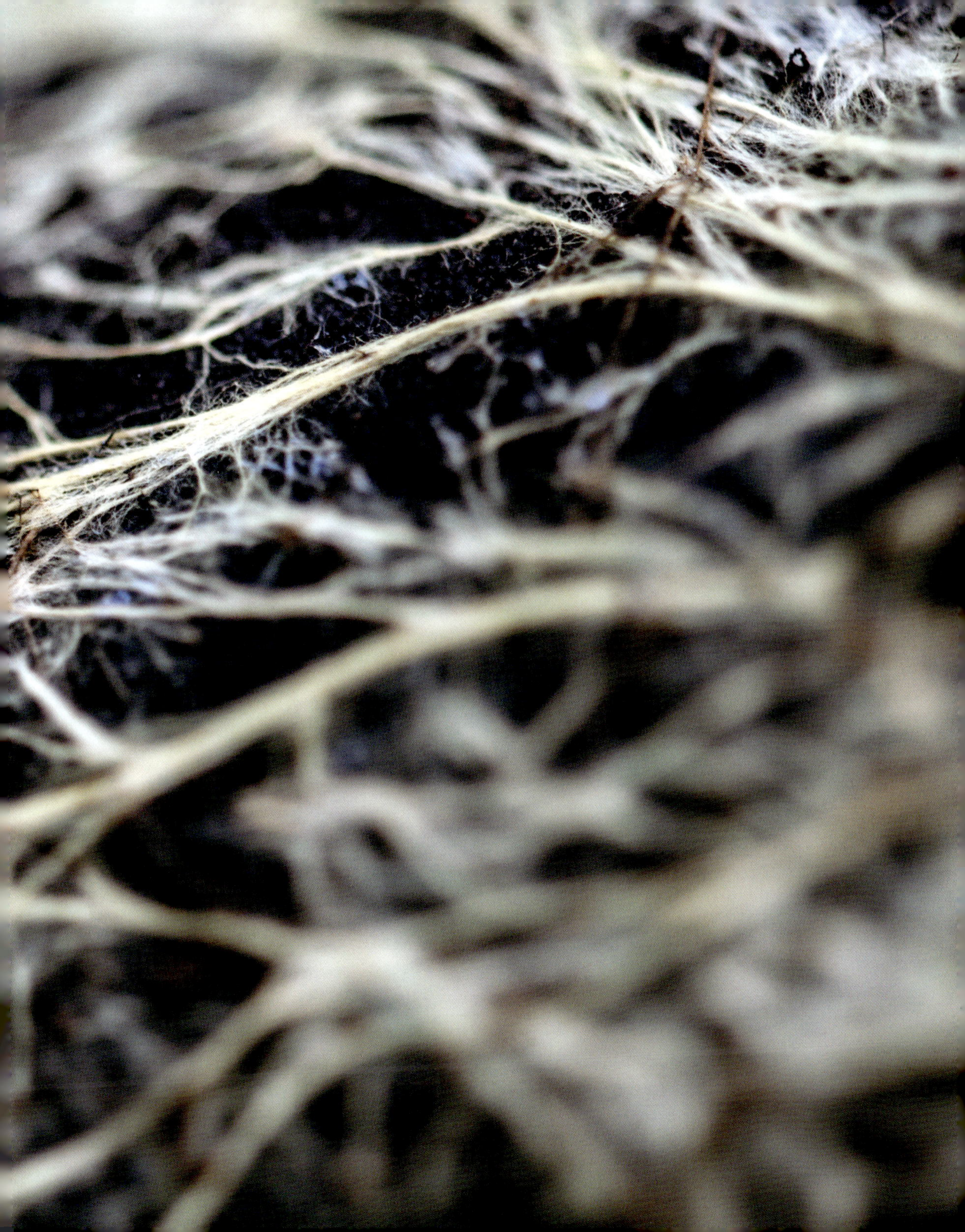

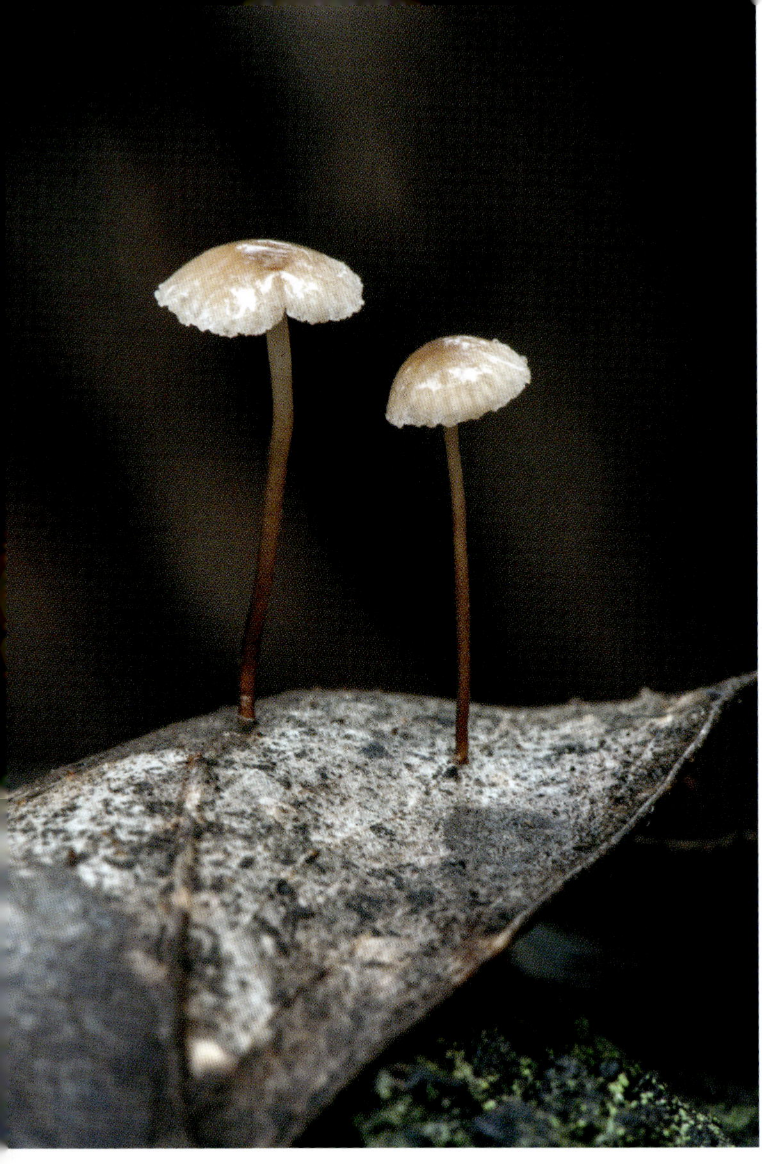

Parachutes (*Marasmius*) on eucalypt leaf, Australia (above)

Mycelium engulfs a eucalypt leaf, Australia (opposite)

Following spread: Whitelaced shank (*Megacollybia platyphylla*), Switzerland

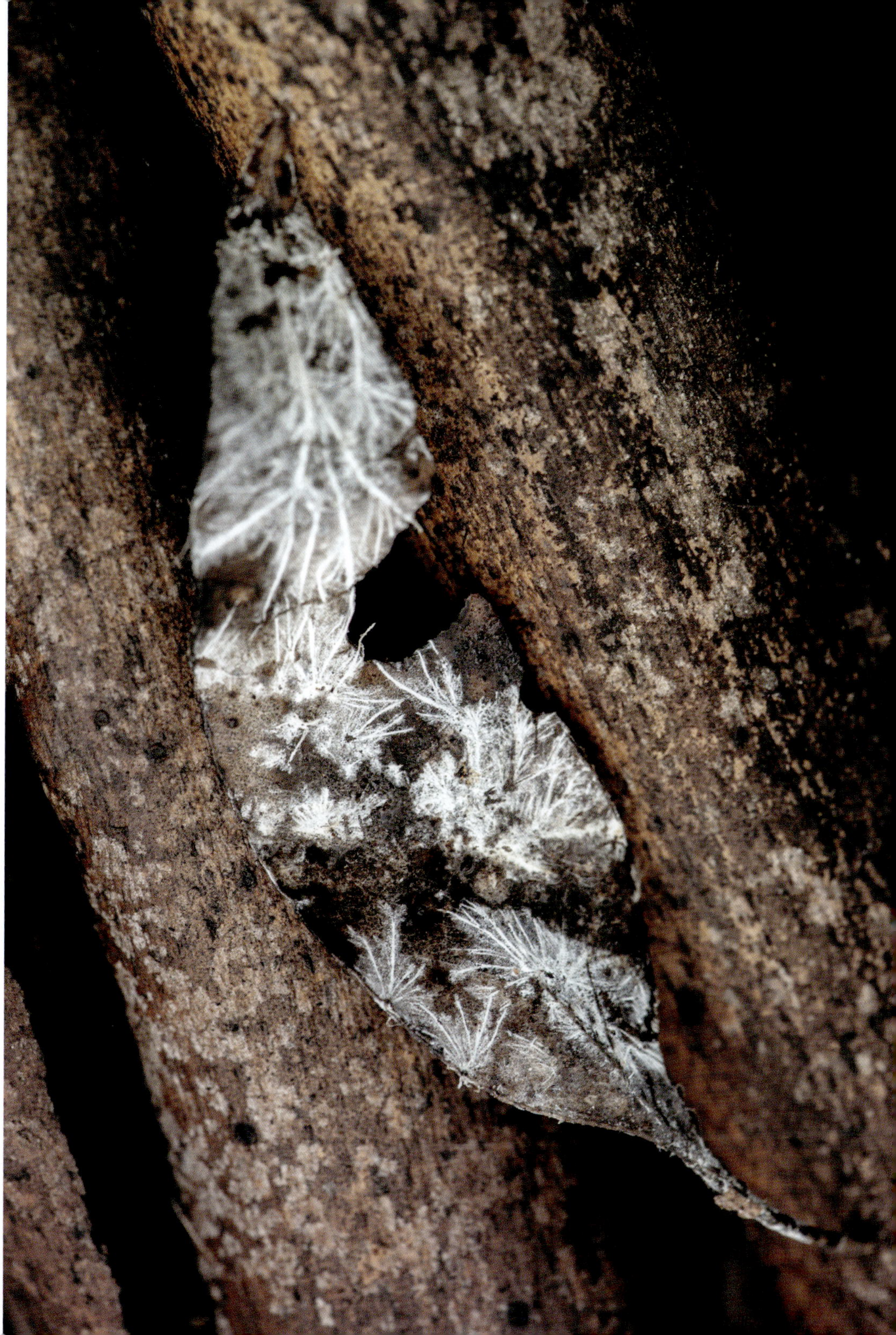

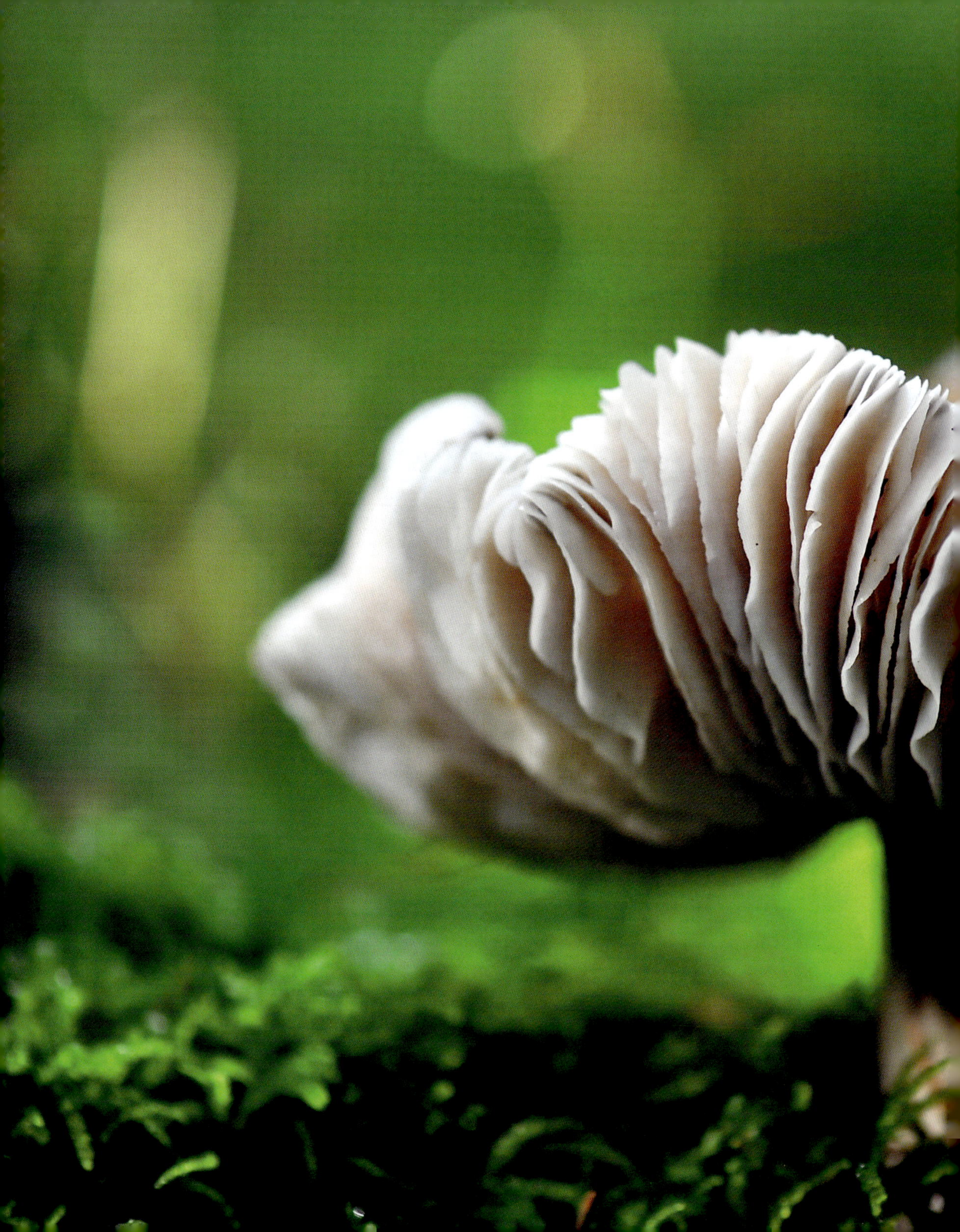

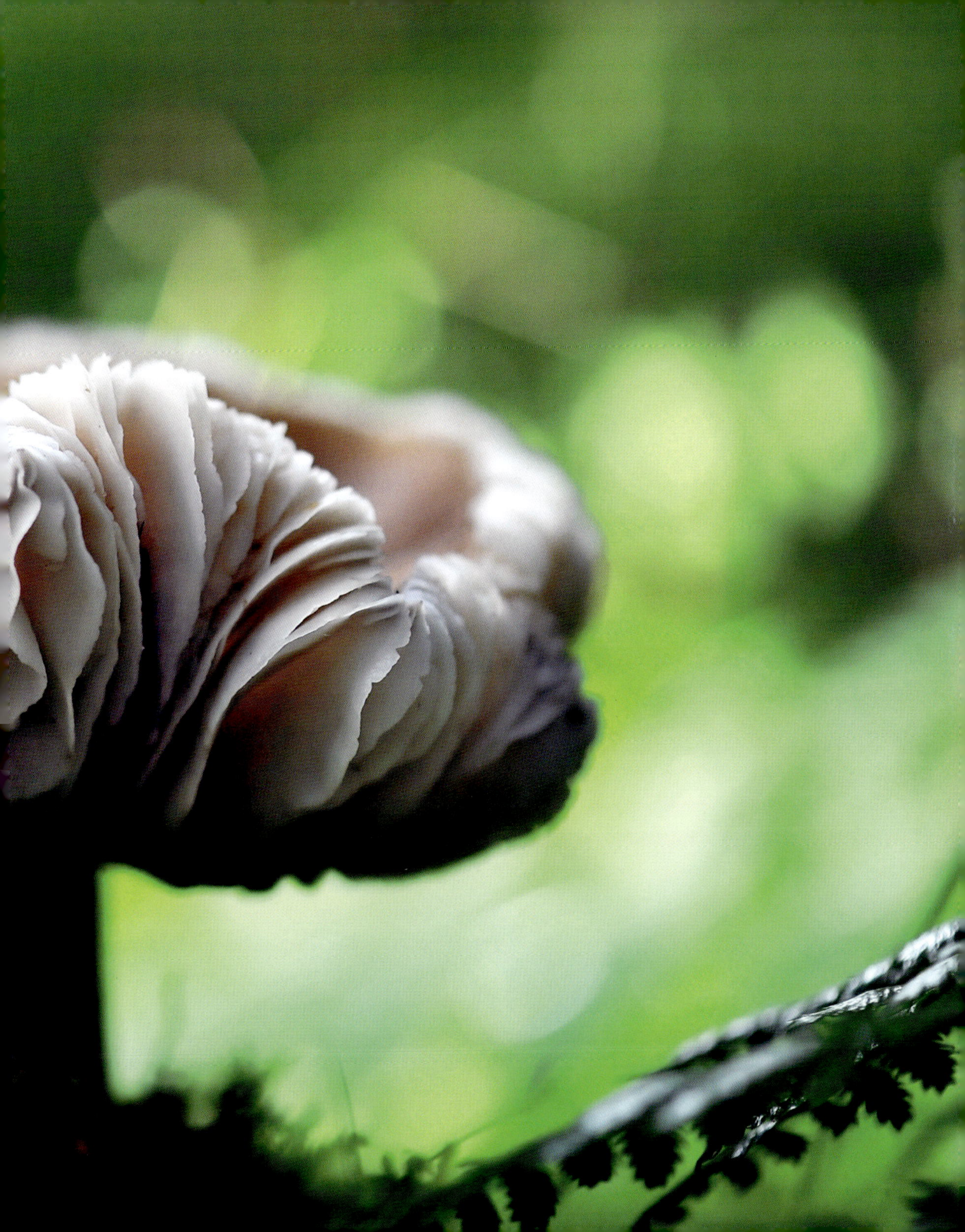

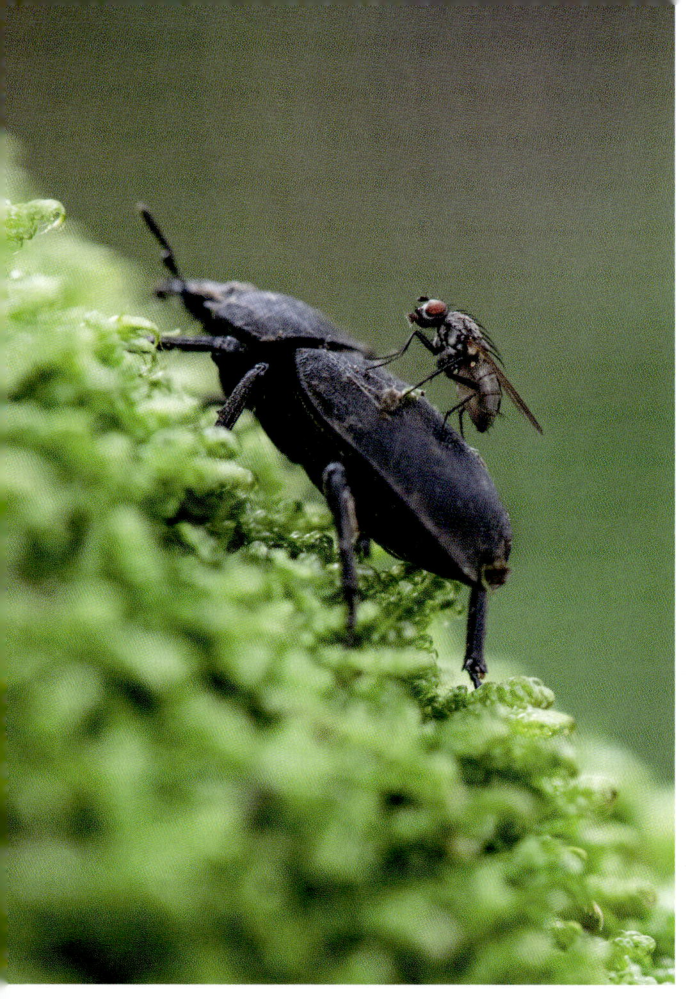

Spore spreaders

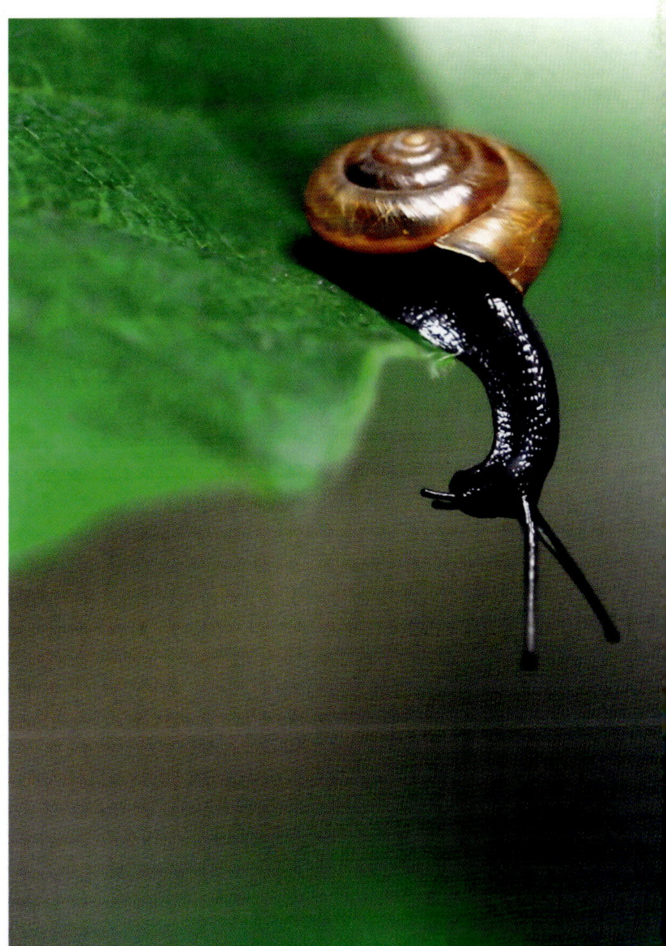

tiny mushrooms appear on the

surface but their mycelium

within the log can be

v a s t

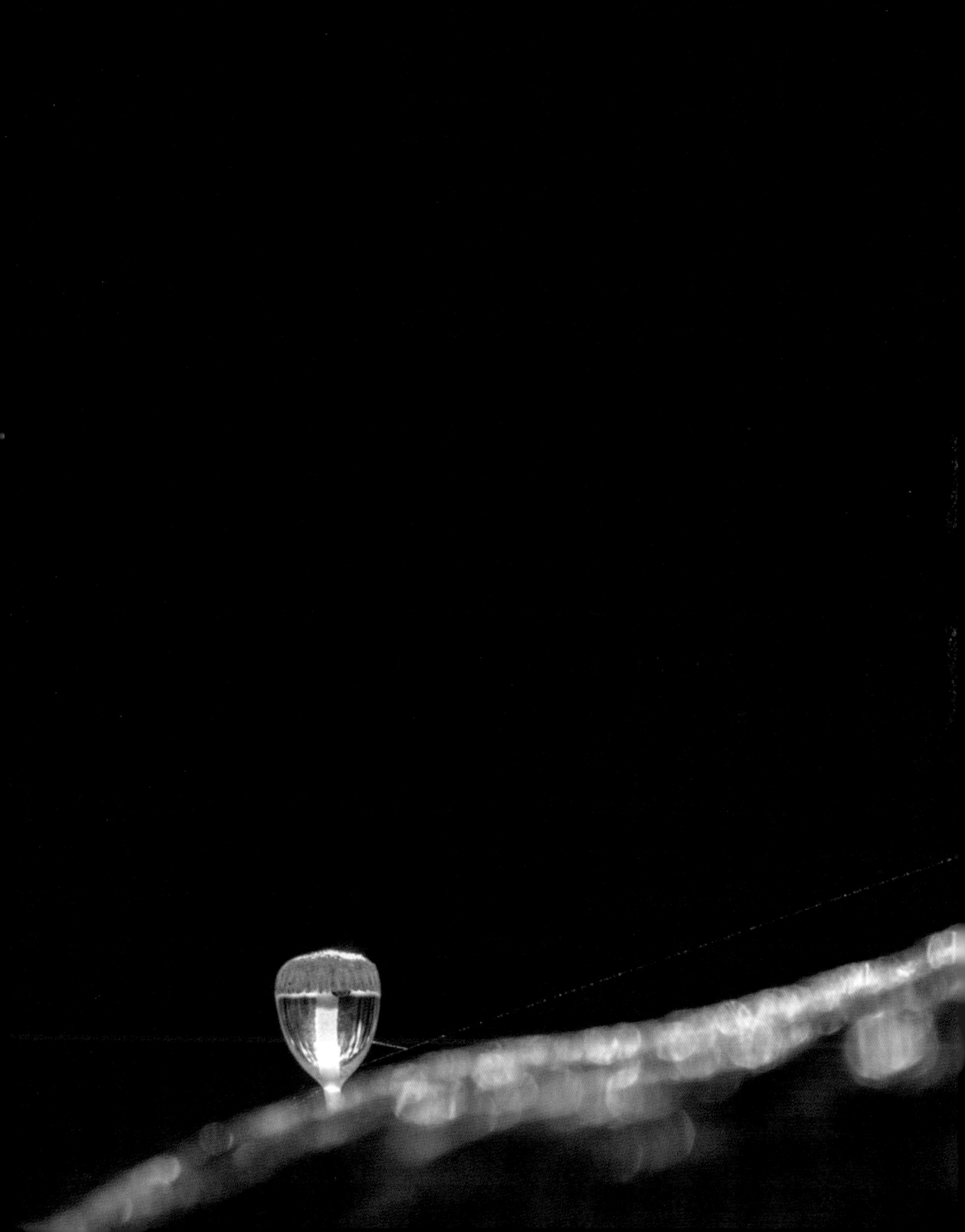

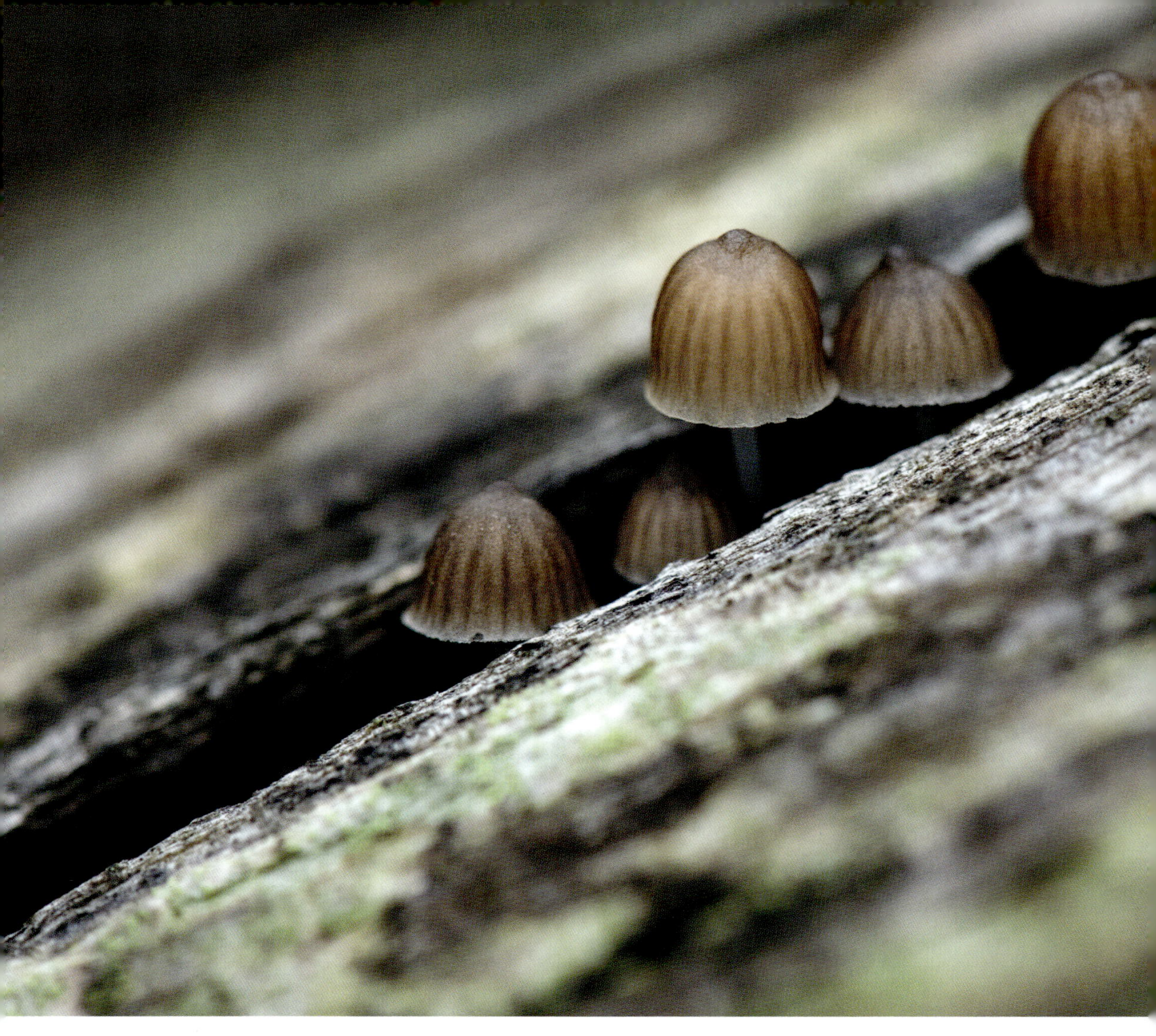

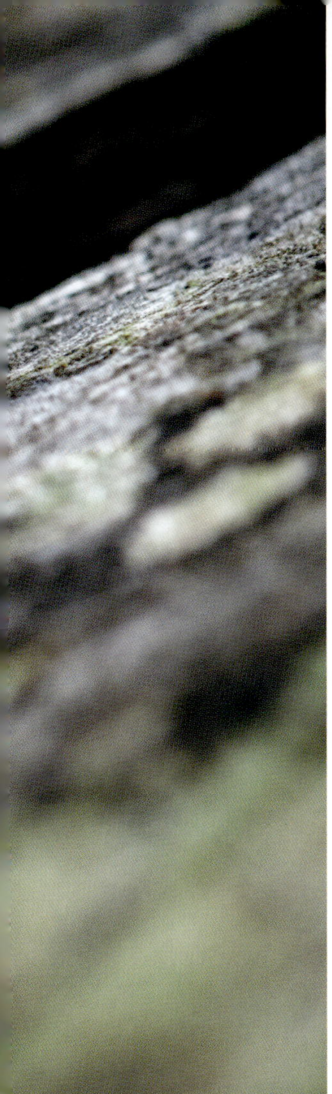

escalating fungi

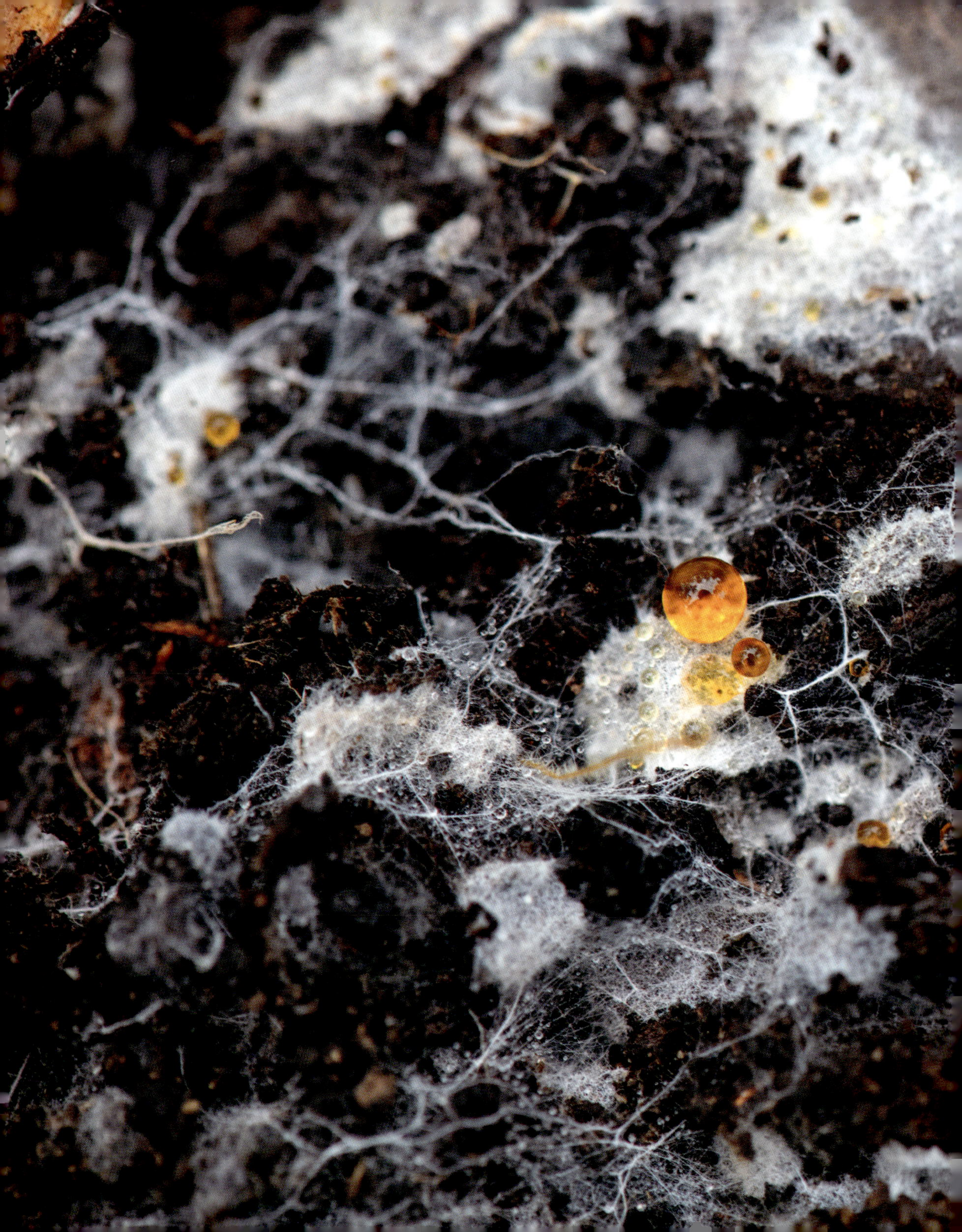

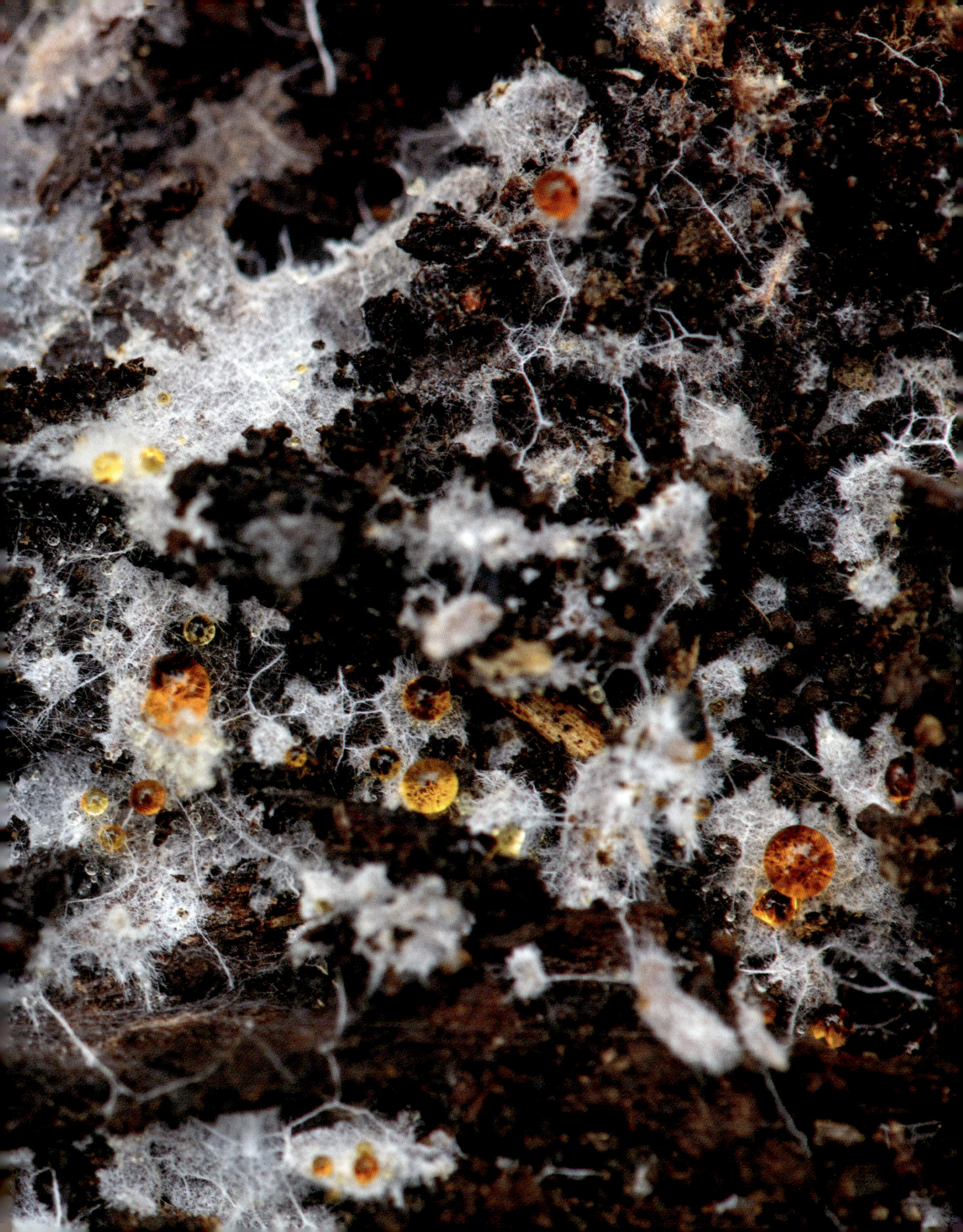

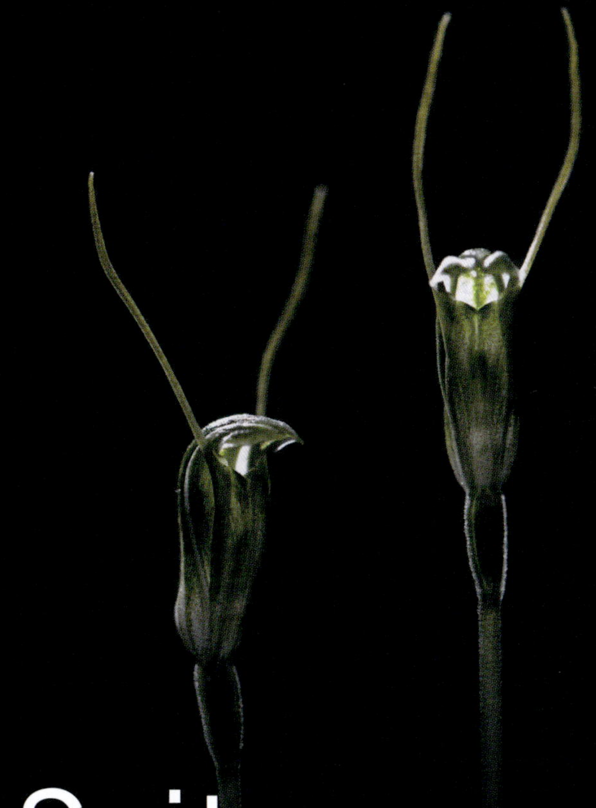

Suitors & assasins

Sanctuaries and reserves for the conservation of species are often located in wild and remote places. I've been fortunate to visit many of these, and experience the fungi and other organisms they harbour. The Tallarook Bushland Reserve, on the other hand, is wedged between the Hume Freeway and the train line that connects the eastern Australian cities of Melbourne and Sydney. It was set up to safeguard the blue finger flower (*Cheiranthera linearis*), a locally significant orchid, and a further 20 orchid species. Early one spring, I wandered around the dry and dusty reserve to the roar of trucks on the freeway and the grind and squeal of freight trains. It wasn't exactly paradise, but there among the weeds and human refuse I found myself entranced by these botanical marvels.

It's easy to see why some people are enamoured with orchids when you observe their intricate details up close. With their unexpected projections, veins and eccentric morphologies, some of which resemble human body parts, they're at once elegant and alien. Many orchids are astonishing imitators: floral 'deception', especially orchid mimicry, is a well-known trick to lure pollinating insects. Some mimic other plants, while others resemble female wasps to fool unsuspecting males into attempting to mate with them. And it's not just about appearance. Some orchids add to their appeal by exuding a pheromonal perfume to smell like the insect they're imitating. But it's when you head underground that orchids become even more enigmatic.

Every orchid you've ever admired, along with the rest of the 28 000-odd species worldwide you might not yet have met, forms relationships known as mycorrhizas with fungi. These are mutually beneficial associations between a fungus and the roots of plants (mycorrhiza = fungus root). Teaming up with other organisms for survival shores up the chances of passing on your genes. Mycorrhizas are among the most ancient alliances. Fungi have been supporting terrestrial plants in this way since those plants first appeared on land. Early plants had minimal root systems and could not extract what they needed from the nutrient-deficient primeval soils. Nearly all plants today form these relationships with fungi.

Although orchids are more recent arrivals, their success as the second-largest family of flowering plants could have a lot to do with these complex collaborations. They first appeared in the northern hemisphere when dinosaurs roamed the land (and some obviously avoided being trodden on) around 83 million years ago. Since then, they've evolved a suite of tricks and talents that has enabled them to spread and colonise a great range of habitats. While some of the largest and most spectacularly beautiful orchids grow in the tropics, many eke out an existence in nutrient-deficient soils such as those at Tallarook.

Orchid seeds are as minuscule as specks of dust and lack endosperm (nourishing tissue around the embryo), so they have no nutrient reserves. That's where fungi help out. Most orchid seeds cannot germinate without a fungal partner, which might be the sole source of nutrition in the first years of an orchid's life. Mycorrhizal fungi provide vital nutrients, especially carbohydrates, that the orchid needs for growth. Although most orchid species, in their adult stage, will develop chlorophyll, photosynthesise and make their own food, many still keep their fungal buddy. The nature of fungus–orchid relationships varies between species, and they will no doubt keep us guessing how each alliance plays out in the secrecy of the subterrain.

Reserves like Tallarook are vital to orchid survival, as many species exist in very restricted geographical areas. If conditions become unfavourable in their region, such as those that could result from climate change, they'll have nowhere else to go. Even slight shifts in temperature or moisture, let alone larger catastrophic events, could well wipe them out.

Hyphae getting hitched

It's a tough slog going it alone, and joining forces increases the chance of survival. Ecosystems, like human societies, build and rely on relationships. The stronger those are, the more robust and persistent the system. Symbioses – arrangements between different organisms that live intimately together – are ancient, and those between fungi and plants could go back more than a billion years. They enable life. The terrestrial biosphere as we know it would not exist if it weren't for the relationships formed between the ancestors of terrestrial plants and fungi.

Although scientists have known about the existence of symbioses for well over a century, these relationships are complex and work in different ways. Humans have created neat categories to differentiate the various relationships that fungi form. Yet some fungi have not read the textbooks that we've written about them, and they blur their relationships and contravene the categories we assign them. Like human relationships, it's hard to know whether these ones are truly mutually beneficial or perhaps a tad parasitic, with one partner benefiting to the disservice of the other. Most relationships, whether human, mycorrhizal or otherwise, play out in different ways at different times in the union. Not all partnerships are equally amicable or sustainable. While the categories inevitably simplify the diverse and complex nature of fungal relationships, they help at least to get us started.

Mycologists split mycorrhizas into two groups based on differences in their anatomy, function and evolution. With ectomycorrhizal fungi, hyphae form an external (ecto-) sheath or mantle around plant roots. Hyphae also form a network that extends between the plant root cells – a bit like mortar surrounding bricks in a wall – called a Hartig net. With endomycorrhizal fungi, hyphae enter and form associations inside (endo-) the cells of plant roots.

Arbuscular mycorrhizas are the most common type of endomycorrhizas. They're so-called because the fungus forms tree-shaped structures, or arbuscules, within plant root cells, where nutrient exchange occurs. These are the oldest and most widespread mycorrhizas, found in more than two-thirds of all land plants.

Although these different mycorrhizas vary physically, they do much the same thing: unite with plant roots and radiate their hyphae into the soil, making nutrients more available, increasing a plant's absorptive uptake of water and nutrients, protecting them from soil pathogens and increasing their resilience to drought. In return, the plant rewards the fungus with sugars produced through photosynthesis.

If you look at the mushrooms growing beneath a tree, you may not only find many individual mushrooms but many different species. Unless you know your fungi well, it's hard to be sure which ones are mycorrhizal (attached to plants' roots) and which are saprotrophs (recyclers) just by looking at the mushrooms. The latter don't attach to plant roots but derive their nutrition directly from surrounding organic matter.

While some mycorrhizal fungi form specific relationships with a particular tree genus, others are generalists and have many different tree partners – and most are probably somewhere in between. Some trees have only a few fungal partners, but most probably unite with hundreds of different fungi.

Because many ectomycorrhizal fungi produce mushrooms or other visible sporing bodies, they are more conspicuous than arbuscular mycorrhizal fungi. Arbuscular mycorrhizal fungi usually produce spores directly from their hyphae – or, if they produce sporing bodies, they're often microscopic and remain underground. Many forest trees form ectomycorrhizal relationships, but there are tree genera, including *Eucalyptus*, that form both types of mycorrhizas. You might be familiar with some ectomycorrhizal gilled fungi that grow with eucalypts like brittlegills (*Russula*) or deceivers (*Laccaria*), or various boletes such as the shaggy cap (*Boletellus emodensis*) or the giant bolete (*Phlebopus marginatus*), or earthballs like the horse dung fungus (*Pisolithus*), or various coral fungi, among many others. All these fungi support the great diversity of *Eucalyptus* tree species that could have hundreds, if not thousands, of different mycorrhizal partners.

Endomycorrhizas also include orchid mycorrhizas, such as those we met at Tallarook, and ericoid mycorrhizas. Ericoid mycorrhizas form between fungi and various berry plants, such as blueberries, cranberries, and heathers. Like arbuscular mycorrhizas, the hyphae penetrate plant root cells, but they form coils rather than arbuscules.

It was in the rainforests of the wild west coast of Vancouver Island, Canada, that I got to know another type of endomycorrhiza – arbutoid mycorrhizas. These form between fungi and madrones (*Arbutus menziesii*) and other trees and hardy shrubs in the family Ericaceae. The madrone is an elegant evergreen with striking, satin-smooth red bark. Its branches shear off in every direction, providing a lovely visual asymmetry among the more regular form of conifers in these forests. Arbutoid mycorrhizas break the rules yet again by sharing features of both ecto- and endomycorrhizas. They're structurally like ectomycorrhizas but their hyphae penetrate the epidermal plant cells and fill them up with hyphal coils. Sometimes they're referred to as a type of ectendomycorrhiza (which sounds a bit like a video game) found in pine, spruce and larch, but they differ in their structure and in the plants and fungi involved.

Another group, monotropoid mycorrhizas, add further to the mycological mayhem by also messing with the rules. These relationships form between fungi and a strange group of plants, among the more common of its members being the insipid-looking ghost pipe (*Monotropa uniflora*). I first encountered these botanical anomalies late one summer among a stand of hemlock on Oregon's Mount Hood. They lack chlorophyll and therefore can't photosynthesise and produce carbohydrates. Although recognised in the 1800s, it took scientists a while to reveal the exact nature of the relationship, but it turns out that the wily ghost pipe is parasitic on a fungus that has a mycorrhizal relationship with a tree. While the tree and the fungus share a mutually beneficial mycorrhizal relationship, the ghost pipe coaxes carbon and nutrients from the fungus and, as far as we know, gives nothing in return. The ghost pipe has somehow fooled the fungus into thinking the relationship has benefits for both. It's a reminder of the incredible complexity of relationships, in this case, a ménage à trois between a parasitic pipe, a mycorrhizal fungus and a photosynthetic tree. It seems the wood wide web of the subterranean might be every bit as prone to freeloading hackers as the world wide web.

Parasites

Not all symbioses are favourable to both partners in a relationship. Parasitic relationships benefit one partner while the other suffers varying degrees of disadvantage, ranging from mild irritation to death. Fungi can parasitise animals, plants and even other fungi. Some fungi, such as honey fungi (*Armillaria*), straddle various modes of existence, moving from parasite to recycler; others form mycorrhizal relationships with certain plants while parasitising others. Some fungi have their cake and eat it – and they will swipe someone else's should the opportunity arise.

Many fungi parasitise plants. Polypores can form large brackets on the sides of trees, while other parasitic fungi, like the famous *Cordyceps* and company, specialise in parasitising a huge range of different arthropods. Other fungi parasitise slime moulds. Stranger still are the myco-parasites that live on or feed on other fungi. The pin mould or bonnet mould fungus (*Spinellus fusiger*) targets bonnet mushrooms of the genus *Mycena* and makes them appear as though they're having a very bad hair day. They'll attack other genera too, including *Amanita*, *Gymnopilus* and *Collybia*, but bonnets seem to be this parasite's popular choice. *Spinellus* means 'little spike', and *fusiger* means 'spindly', and indeed the fungus looks like a spiky, spindly pile of pins. How do these fungi find their fungal hosts? It's not known for sure, but insects and other arthropods are likely to be vectors, especially given the great range of these tiny creatures that inhabit the host fungi. While parasites rarely rate highly on most people's lists of favourite organisms, they're vital to ecosystem regulation and stability. Without these critical components of the complex web of relationships, forests would fail to function. Next time you encounter a parasite, fungal or otherwise, be sure to let it know what a fabulous job it's doing in helping to keep the planet turning.

Following spread:
Dwarf greenhood orchid (*Pterostylis nana*), Tallarook Bushland Reserve, Australia

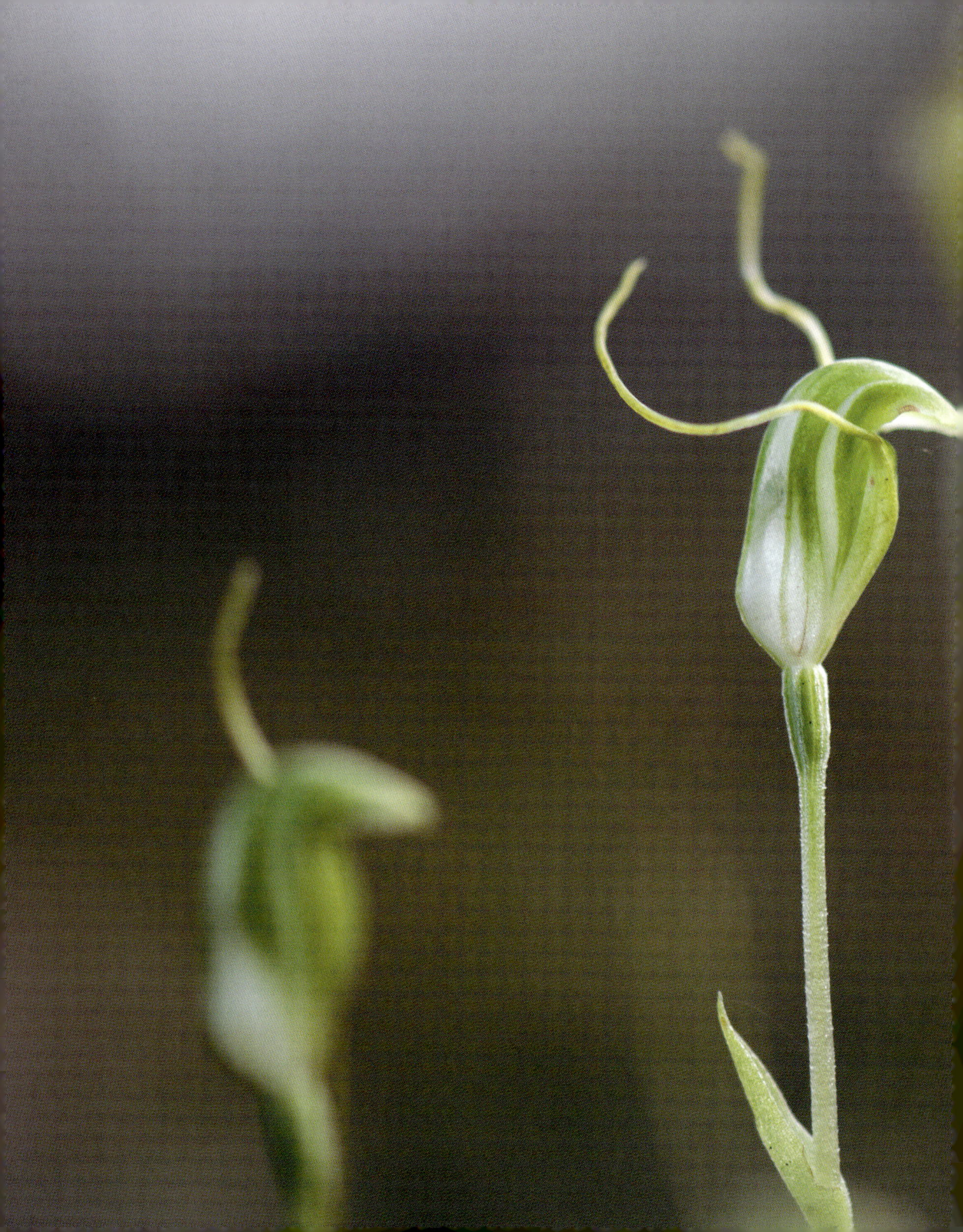

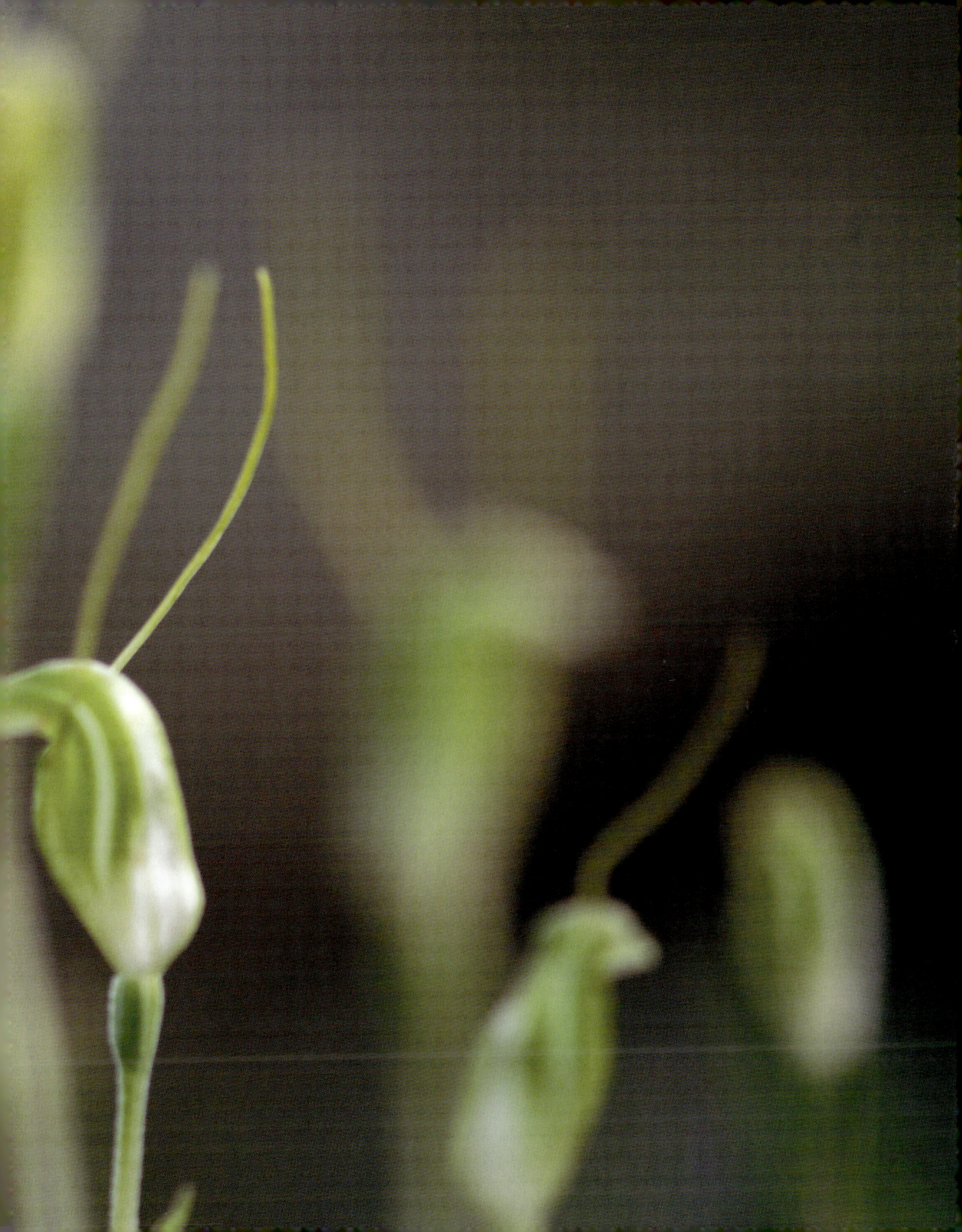

Left:
Dwarf greenhood orchid (*Pterostylis nana*), Tallarook Bushland Reserve, Australia

Right:
Waxlip orchid (*Glossodia major*), Tallarook Bushland Reserve, Australia

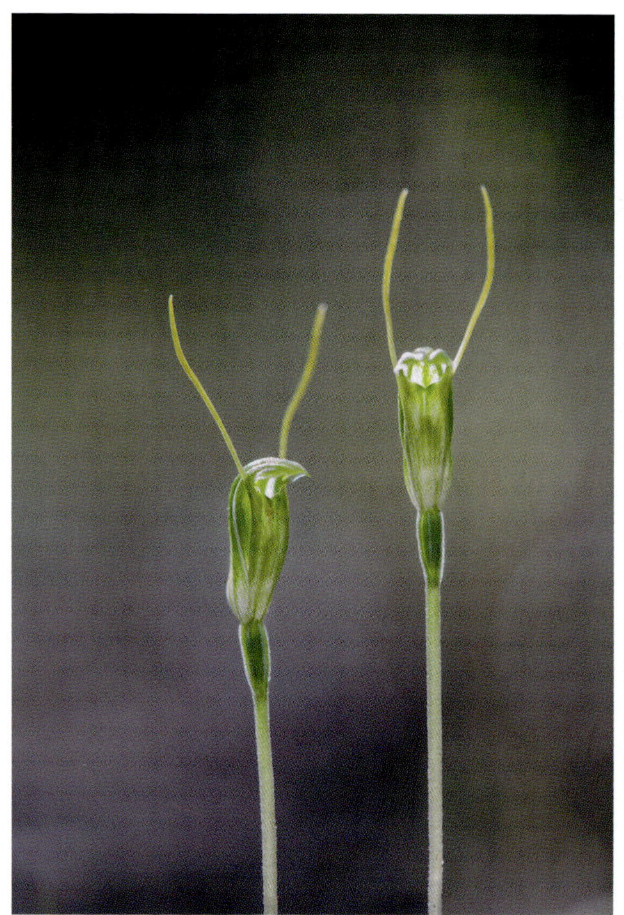

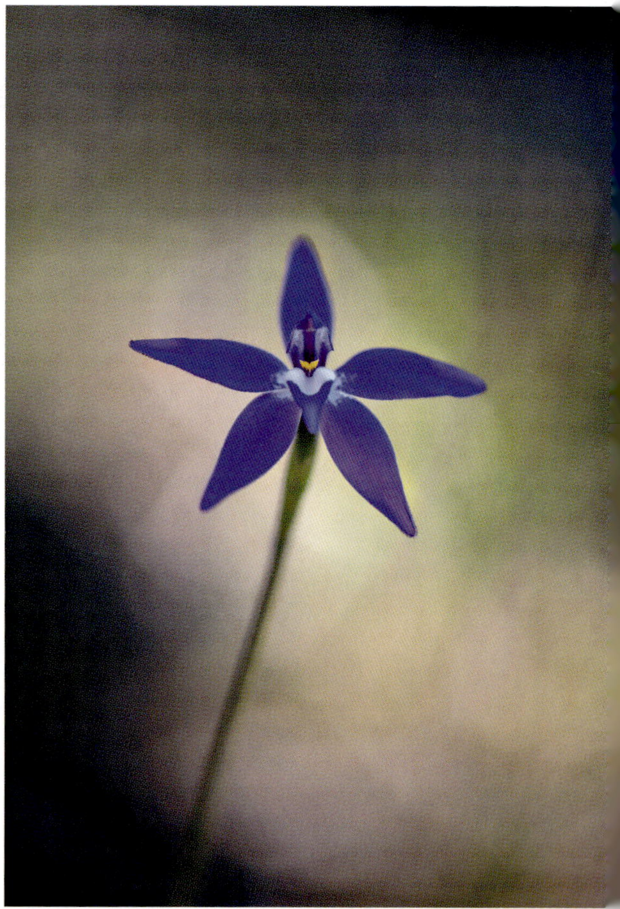

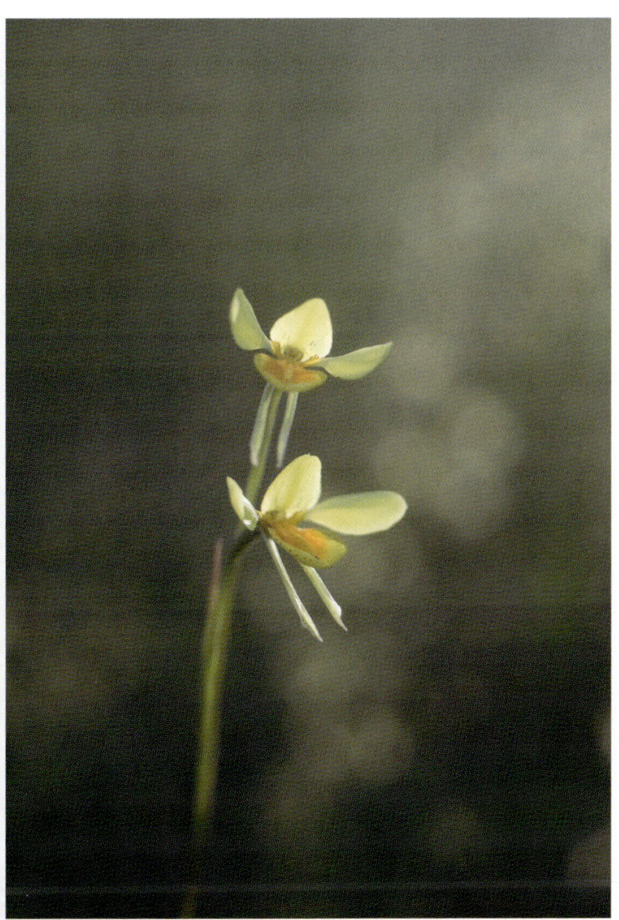

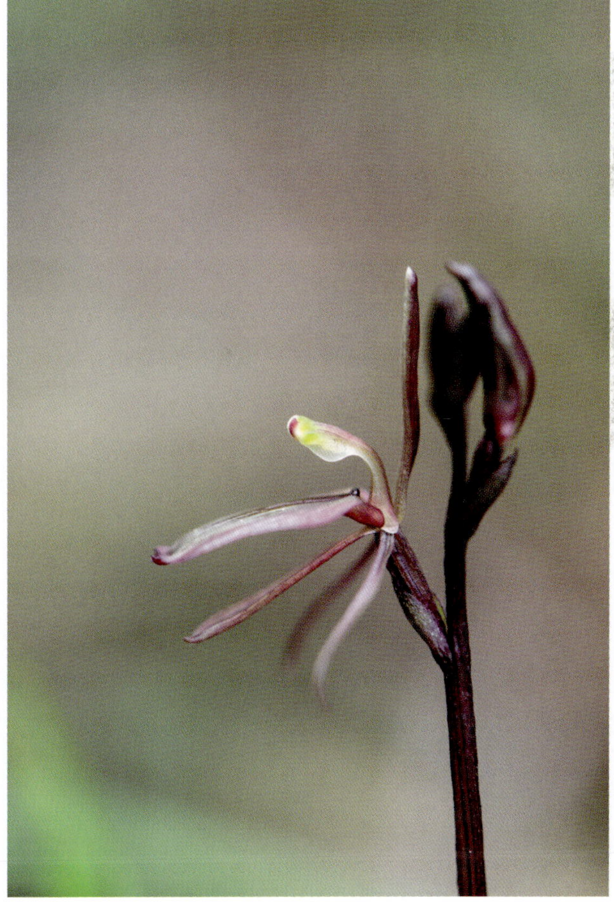

Left:
Golden moth orchid (*Diuris chryseopsis*), Colin Officer Flora Reserve, Australia

Right:
Gnat orchid (*Cyrtostylis reniformis*), Tallarook Bushland Reserve, Australia

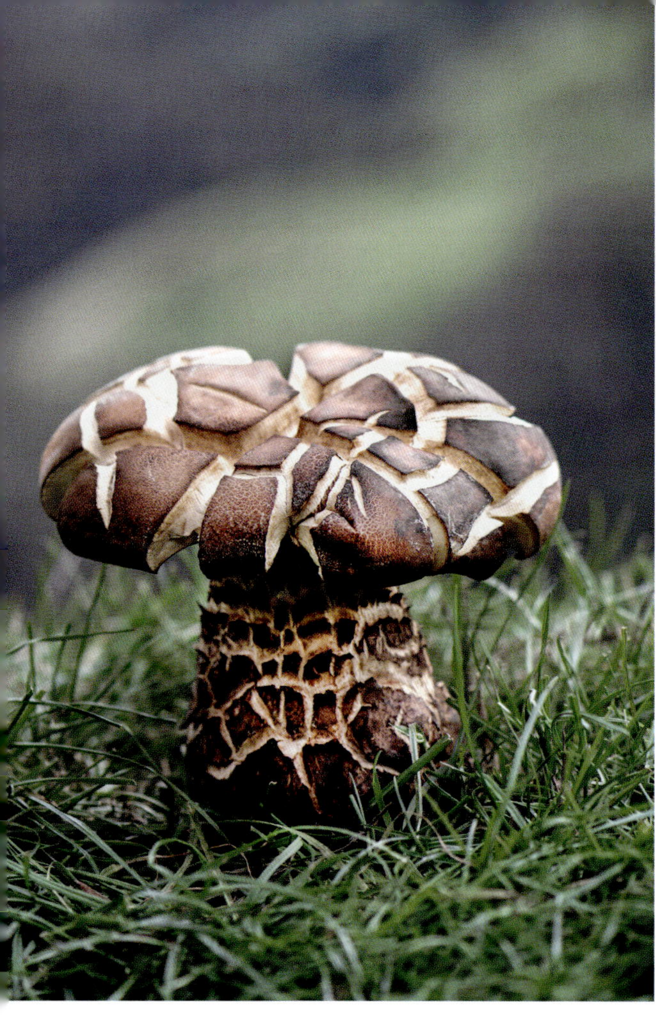

Many fungi form ectomycorrhizal relationships with eucalypts, including the giant bolete (*Phlebopus marginatus*) (above) and the shaggy cap (*Boletellus emodensis*) (opposite)

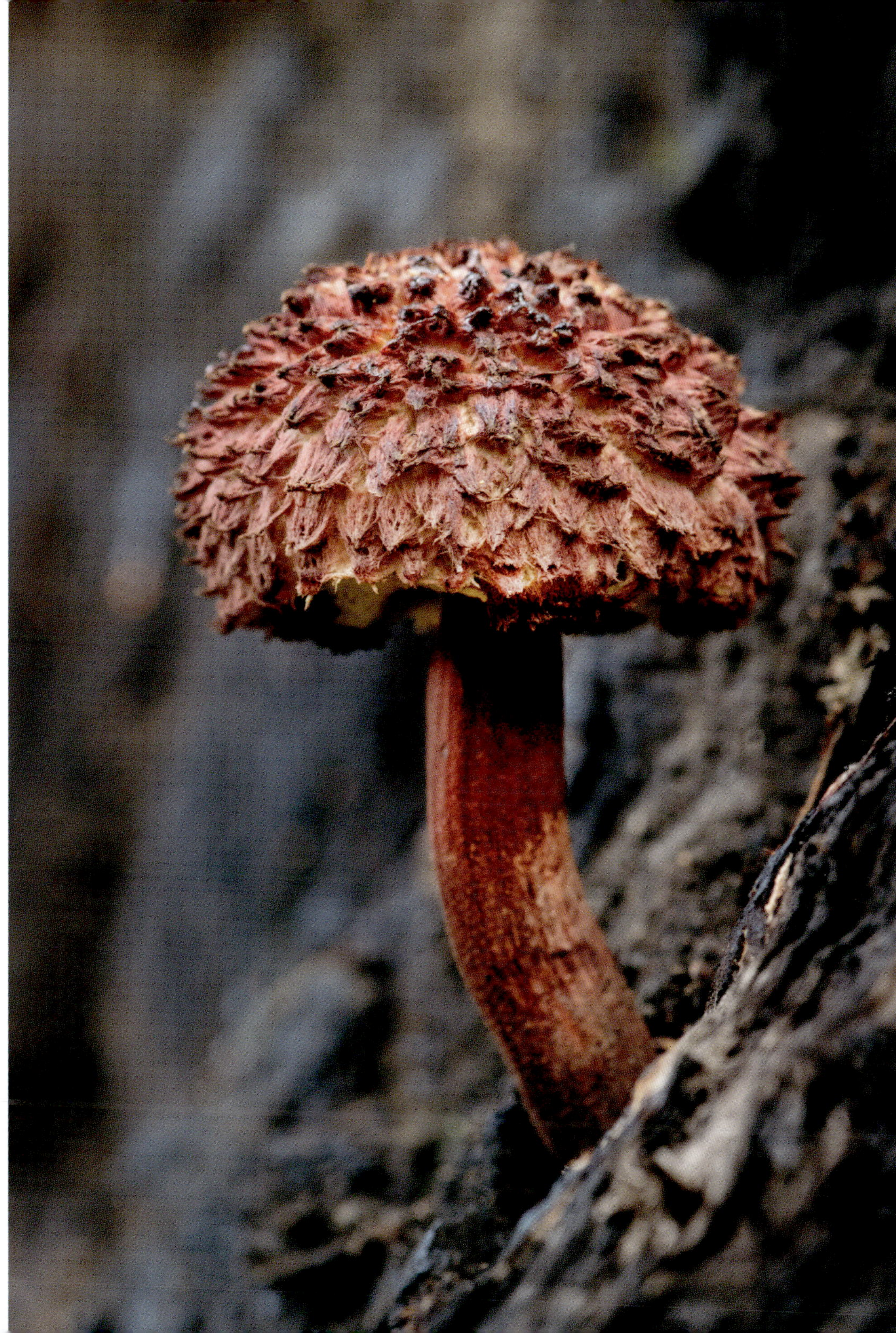

Horse dung fungus
(*Pisolithus marmoratus*)
in cross-section, Australia

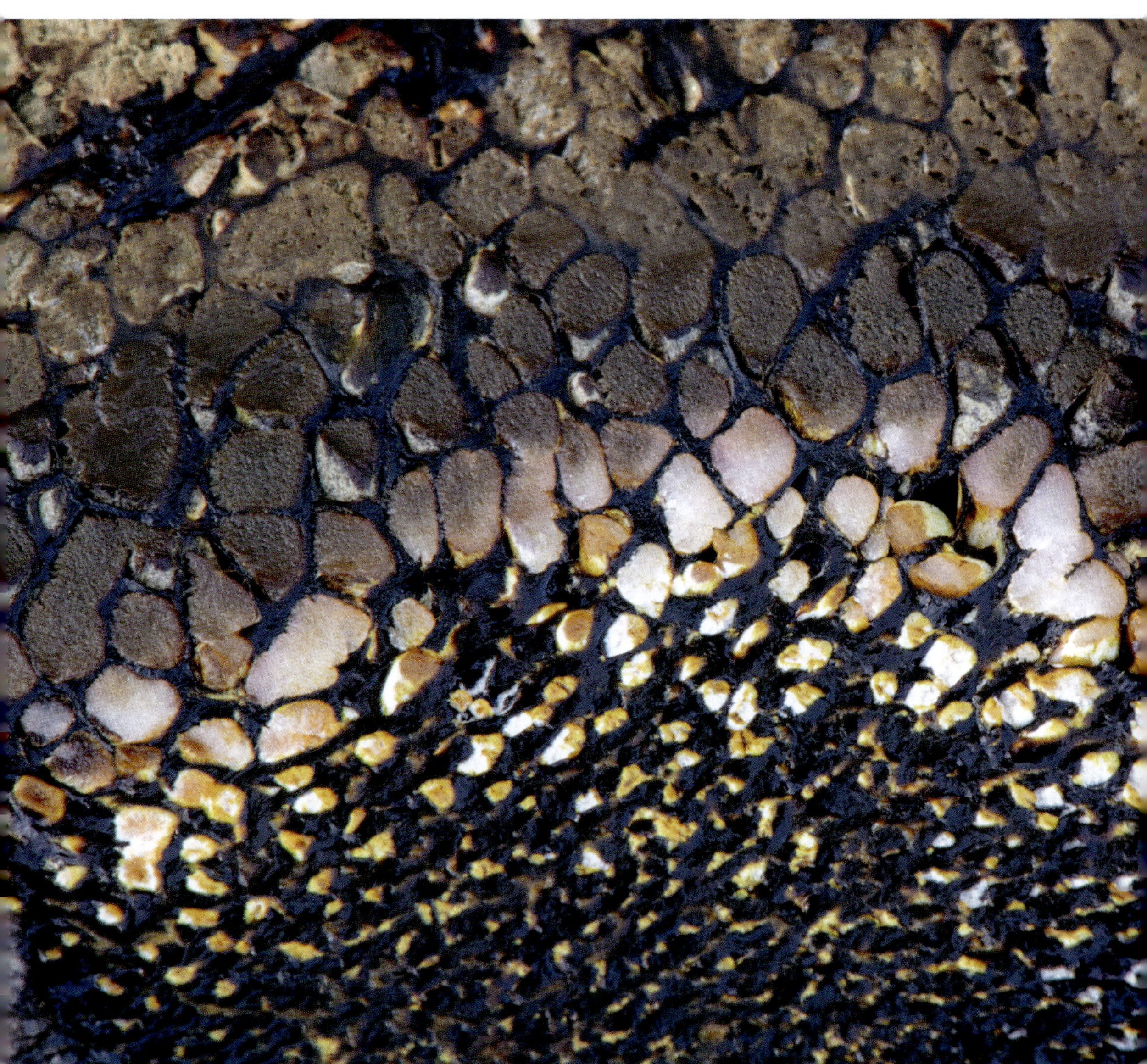

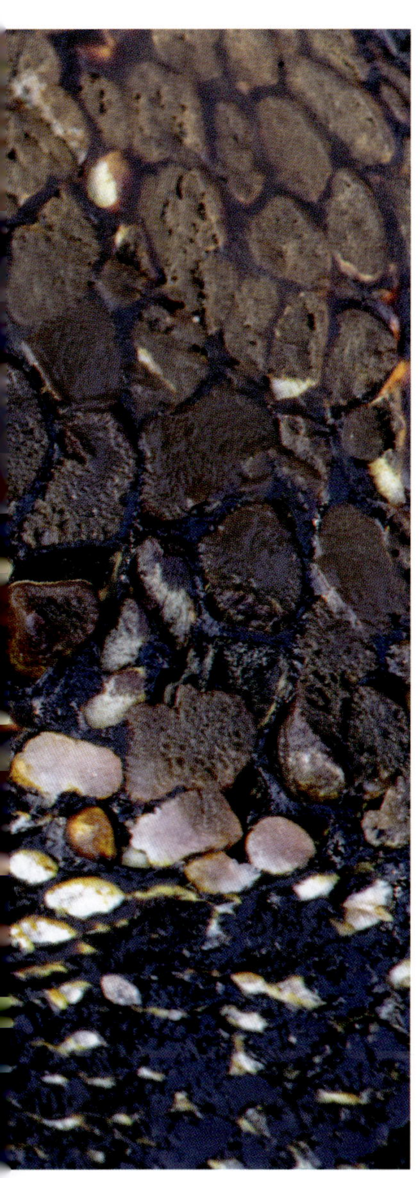

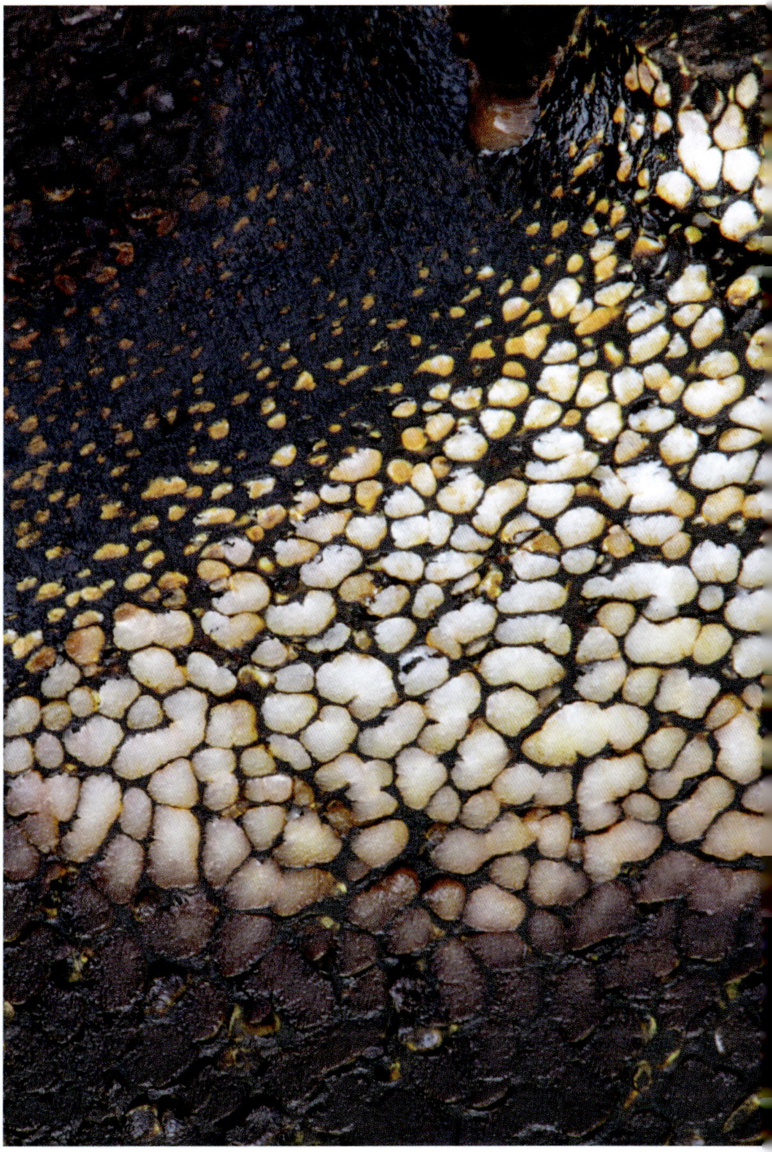

Parasitic bonnet mould fungus (*Spinellus fusiger*), Germany

FUNGI that eat FUNGI

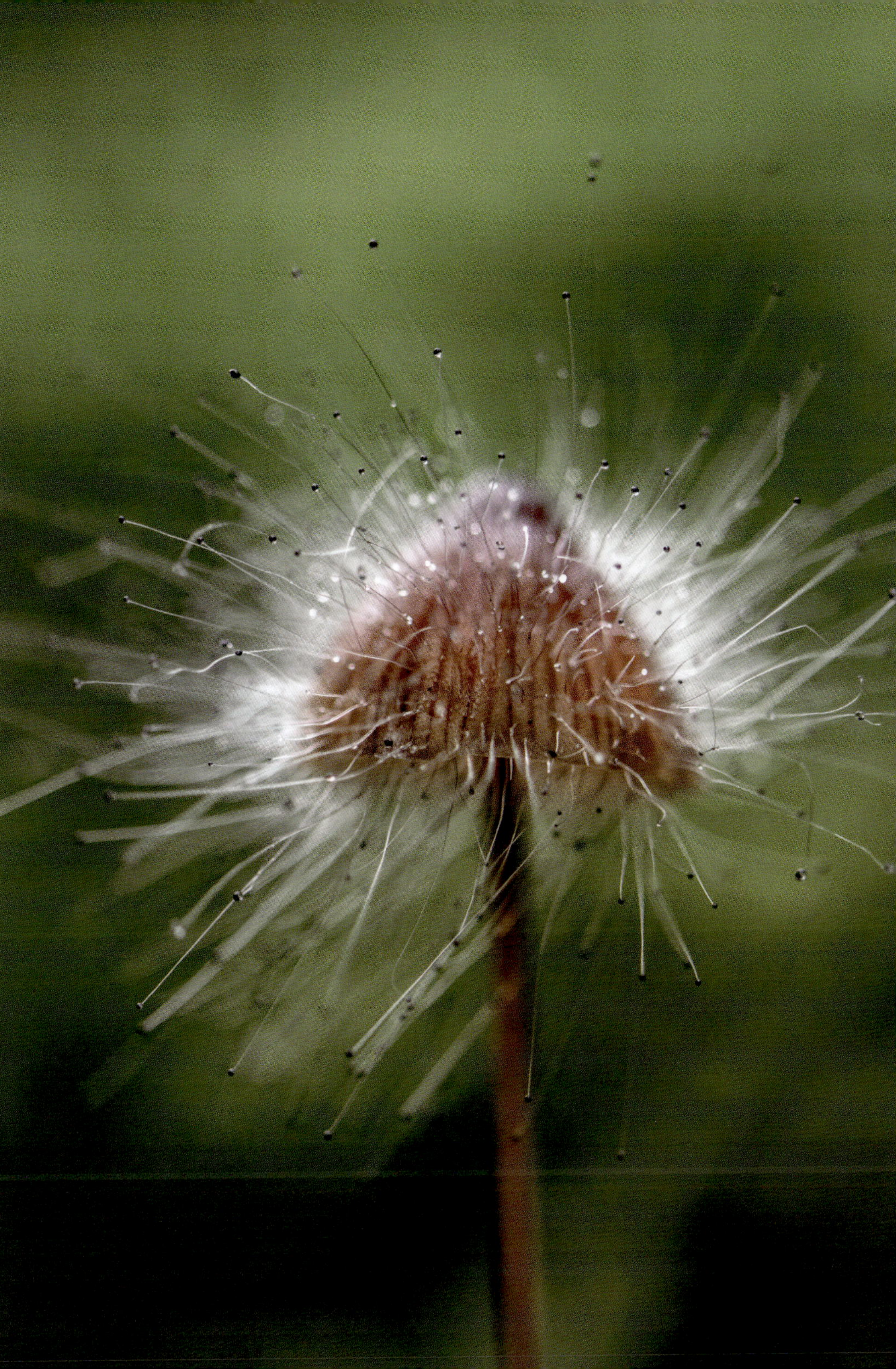

Parasols & oddballs

From the Lac de Bienne at Vingelz in Switzerland, you can just make out the snowy peaks of the Bernese Alps. One coolish early autumn morning, as mists rose off the water and the first rays of sunshine glanced off the glassy surface, I swam alongside mushroom inspector Barbara Thüler. We'd been in the water for about half an hour and Barb was showing no sign of turning back. I began to wonder if she intended on swimming the entire 15-kilometre length of the lake when she paused mid-stroke, raised a wet eyebrow, and questioned, 'Sulphureus?'

'What?' I asked, thinking of *sulphureus* as the epithet of various fungi such as *Laetiporus sulphureus*, or chicken of the woods, but right then we were swimming in the lake, not roaming the forest in search of fungi. Sure enough, Barb tilted her head and nodded toward the shore where the sulphur-coloured shelves of this large, distinctive fungus protruded from the stump of a beaver-felled willow. Despite being a good 50 metres away, it stood out like a beacon, at least to a mushroom inspector. I then realised that mushroom inspectors have no 'off switch', especially in autumn when fungi fill the forests and our consciousness, and that Barb constantly scans the environment for mushrooms, even while swimming across a lake.

To make sense of the great diversity of fungi, it helps to know who's who. As a mushroom inspector, Barbara is more familiar than most with the great variety of reproductive manifestations into which a fungus can wrangle its hyphae. She's one of more than 1400 inspectors in Switzerland whose job it is to identify foragers' collected fungi, in particular to differentiate edible mushrooms from toxic Doppelgänger. Although mycologists today identify fungi largely by their genetic material, if you're a forager or photographer (or a swimmer), recognising fungi by their form – their morphology – is often more convenient than convincing a mushroom to surrender some of its genes.

Like humans, fungi are astonishingly diverse in how they appear. Many fungi are discreet, while some are flamboyant, even voluptuous. Others appear disfigured or grotesque. Some make a fuss, ejecting their spores with much drama or stench. Most are small and brown and drab. Yet all are remarkable in their own way and often conceal compelling backstories. It's an absolute wonder how certain sets of conditions and evolutionary pressures align to bring all these forms into existence.

Calling fungi by names makes it easier to talk about them. However, telling mushrooms apart by their appearance can be tricky. Different species can appear deceptively similar, while other individuals within a species can be so variable in their phenotype – that is, how they appear, their observable traits – as to look like different species. Distinguishing between taxonomically significant variations and those caused by a hailstorm, prolonged dry spell or other environmental influences adds to the challenge. The candlesnuff fungus (*Xylaria hypoxylon*) that pops up on fallen wood can look like a tiny club, a grasping hand, antlers or antennae, among other quirky forms. Fungi are remarkably plastic, and many can switch growth forms in response to different environmental stimuli. For example, a bracket fungus growing on a tree trunk that then falls to the ground can reorient itself 90 degrees so that its spore-producing tubes are held vertically, and spores can continue to be released from its pores. The other species that share its space, and with which it cooperates, competes or interacts in some way, also influence fungus morphology. Yet all the various fungal configurations and subtle differences in their parts allow us to differentiate a good deal of them.

On spore curviness

Mushrooms and other sporing bodies are simple structures when compared to, say, a flower, but they vary in design in unimaginable ways. All of them serve the same sole purpose: releasing spores. Hundreds of millions of years of evolution have produced some remarkable morphologies and strategies for spore dispersal. Although artists and scientists have considered the shapes and colours of flowers for centuries, mycologists have only recently started making meaning from spores and the morphology of sporing bodies.

Umbrella-shaped mushrooms are an ingenious way to protect spores from getting wet, and this form has evolved multiple times throughout evolutionary history. If you peer beneath the caps of different mushrooms, you'll notice wildly varying configurations of their fertile surfaces, also known as hymenia. Many have radiating thin blades, or lamellae. Some have shorter lamellae in between the longer ones, known as lamellulae, or are bifurcate (split in two). Other mushrooms have pores at the end of tubes that vary in colour, size, shape and arrangement. Mushrooms like the hedgehog fungus have tooth-like projections. Chanterelles have 'skin folds' that radiate from the stipe like lamellae but are blunter and more irregular, and often fork and fuse or have ribbing. The varying forms have likely evolved to maximise the spore-producing surface area and protect spores until they're ripe and ready to be released.

Despite their microscopic size, spores themselves are remarkably variable in shape, size, and surface architecture. Some fungi have large spores, while for others it's a numbers game of producing vast numbers of smaller spores. Smaller spores generally disperse further than larger ones, and even more so if they're discharged from a tall mushroom with a higher point of spore release. But larger (or very long) spores are often more resilient and less prone to ultraviolet radiation and desiccation (because of their smaller surface-to-volume ratio), which could explain why larger-spored mushrooms often appear earlier in autumn when it's warmer.

The spores of many species have distinct and variable patterns of ornamentation: some are ridged or reticulate, while others are warty or spiny. Spores also come in a range of shapes with delightful descriptors such as globose (globelike), ellipsoid, cylindrical, elongated, sigmoid (S-shaped), allantoid (sausage-shaped), straight or irregularly curved. Trying to make sense of why they manifest in all these shapes has mycologists scratching their heads. A group of Nordic mycologists notes that 'very little is known about the meaning of curviness of spores', but they suggest it relates to the way they attach to surfaces or are dispersed by animal vectors. The conundrums that keep some scientists awake at night!

These mycologists have also found a connection between spore morphology and trophic guild – that is, whether a fungus is mycorrhizal, saprotrophic or parasitic. The researchers thought this could be linked to the different destinations that various spore types need to reach. Mycorrhizal fungi, for example, usually have larger and more ornamented spores than saprotrophic fungi. They must ensure their spores can germinate next to the root tips of the plants they unite with beneath the soil surface, and they might need an arthropod or two to help get them down there. Spores ornamented with bumps or spines could attach more readily to invertebrates than smoother spores. Knobbly spores have been observed on the bodies of mites, for example. While mites don't walk miles, long-legged or more mobile mite-eating creatures help get spores further afield, as some spores can survive the journey through digestive tracts. However, we still don't know much about all the other differently decorated spores and their carriers, or how spore architecture, size and shape affect wind or water dispersal.

While many fungi rely on wind, some fungi package their spores in a stinky trick or delectable bribe – in the form of stinkhorns or truffles – to entice animals to assist in the spread of spores. Others, such as bird's nest fungi, harness the energy of a raindrop to fling out their spores. Some fungi that live in extreme conditions have special thick-walled spores called chlamydospores, which are typically darkly pigmented and rich in lipids that allow them to lie dormant until conditions improve for germination. Whichever way fungi get their spores out into the world, their impressive diversity and ubiquity suggests they've got it mastered.

Spectrum

The rainbow fungus (*Trametes versicolor*), as it's known in Australia (or turkey tail to the Americans), is so-called because of its delightful spectrum of colours from greys, greens and blues through to browns, reds, amber and beyond. German speakers call it *Schmetterlingstramete* or butterfly mushroom, also a reference to its colourfulness. Its species epithet *versicolor* means many colours. This common wood-recycling fungus grows throughout the world and several Asian cultures revere it for its alleged medicinal qualities.

Homo sapiens have evolved to notice colour. It's usually the first thing we comment on when we spot a mushroom. Many fungi produce brightly coloured sporing bodies. Waxcaps (*Hygrocybe* and company) appear in a kaleidoscopic range of colours, but other genera like *Cortinarius* and *Russula*, and the coral genus *Ramaria*, also sport some stunningly colourful species. Some fungi, boletes especially, delight with a dramatic colour change when their 'flesh' is bruised. The colour of a mushroom often changes as a natural part of the developmental process or with exposure to wind, sun or rain. A fly agaric, for example, can range from fire-engine red to orange, yellow, lemon, and even white depending on different environmental conditions. The whites of the ivory waxy cap (*Hygrophorus eburneus*) or porcelain fungus (*Oudemansiella mucida*) are very different to the 'shades' (hue and saturation) of white *Agaricus* mushrooms. These subtle differences result not just from the mushroom's pigments but also its translucence, its texture and our individual ability to detect certain tones. There's likely to also be a genetic or physiological basis for colour variations that is not yet understood.

Among the extraordinary range of fungal pigments are melanins, which provide brown to black colours (e.g. wood ears), carotenoids (e.g. golden chanterelles), flavins, indigo, azaphilones, anthocyanins (e.g. the striking purple of the amethyst deceiver), quinones and phenazines. Most knowledge about fungal pigments comes from microfungi like *Aspergillus* and *Penicillium*, fungi that are familiar to those in the medical, brewing and baking industries. Other fungi, especially lichens, also produce pigments, which have been used for centuries as natural dyes.

Whether there are ecological or evolutionary advantages for a mushroom to be a particular colour is not known. With plants and animals, pigments are involved in biological processes including camouflage, attracting pollinators or partners, protection from ultraviolet (UV) light and the use of solar energy for metabolism. But is there anything *meaningful* we can derive from a mushroom being, say, purple rather than red? The development of pigments could be an adaptive means of survival to protect cells from environmental stresses, but it may also play a role in ecological interactions with other organisms. For example, carotenoids provide UV protection, while melanins help protect mushrooms from radiation, temperature extremes and desiccation. Melanins also have high-tensile strength, and provide structural rigidity and stability in cell walls. Darkly pigmented melanised spores can remain viable in soils for longer than those that are hyaline (colourless).

The shade of an organism – its lightness or darkness – affects how it absorbs heat. If you've ever found yourself in a black car without air conditioning in a traffic jam in the height of an Australian summer, this idea will be familiar to you. The theory of thermal melanism predicts that darker colouration is beneficial to ectothermic (cold-blooded) animals in colder regions. For example, darker assemblages of insects often occur in environments with high UV radiation. But what about fungi? A group of European researchers found that the theory translates across kingdoms, with mushrooms being darker in colder rather than warmer climates. Because dark-coloured mushrooms heat more rapidly than light-coloured mushrooms, they could have greater reproductive success in cold environments. Yet we still don't know if warm mushrooms release viable spores more effectively.

Mycologists are trying to fathom what these findings mean for fungi in a changing climate. Could a warmer environment favour lighter-coloured species and threaten darker-coloured ones if they push beyond their physiological limits and overheat? Such shifts have been reported in insects, but insects respond more quickly than fungi to environmental changes because of their greater mobility. Comparing a whole organism (insect) to an ephemeral part of an organism (like a mushroom) also has obvious limitations. At this point, it's another fungal mystery to unravel.

As a photographer, you become finely tuned to light and colour. But sitting on the forest floor, photographing these marvellous organisms and feeling like hypothermia is about to set in (despite my black insulated jacket), I can't help wishing that the thermal melanism theory applied to warm-blooded organisms too.

The various designs of sporing bodies serve one purpose: releasing spores. Left: coral tooth fungus (*Hericium coralloides*), New Zealand; right: blistered cup (*Peziza vesiculosa*), USA; opposite: white saddle (*Helvella crispa*), Finland

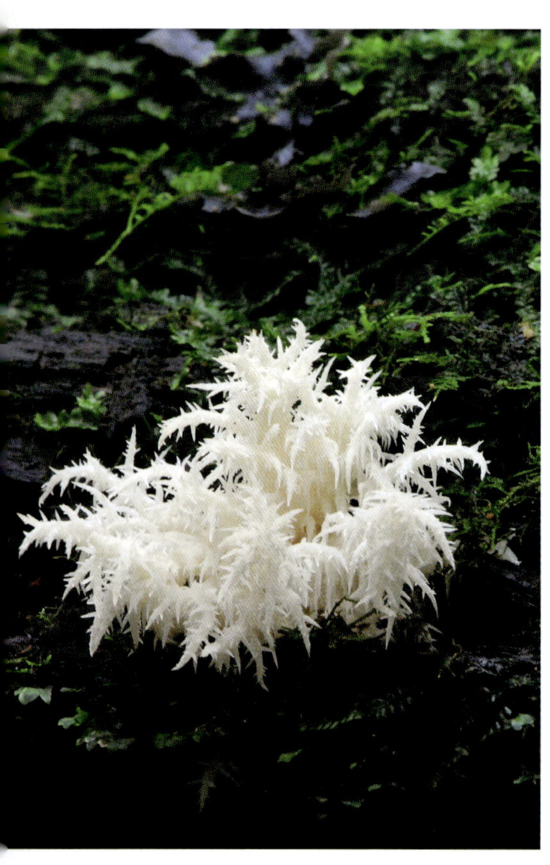

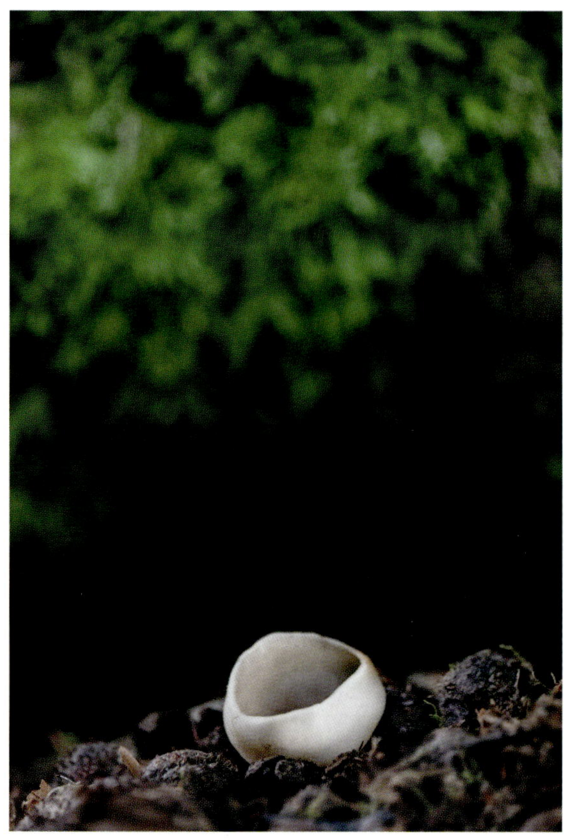

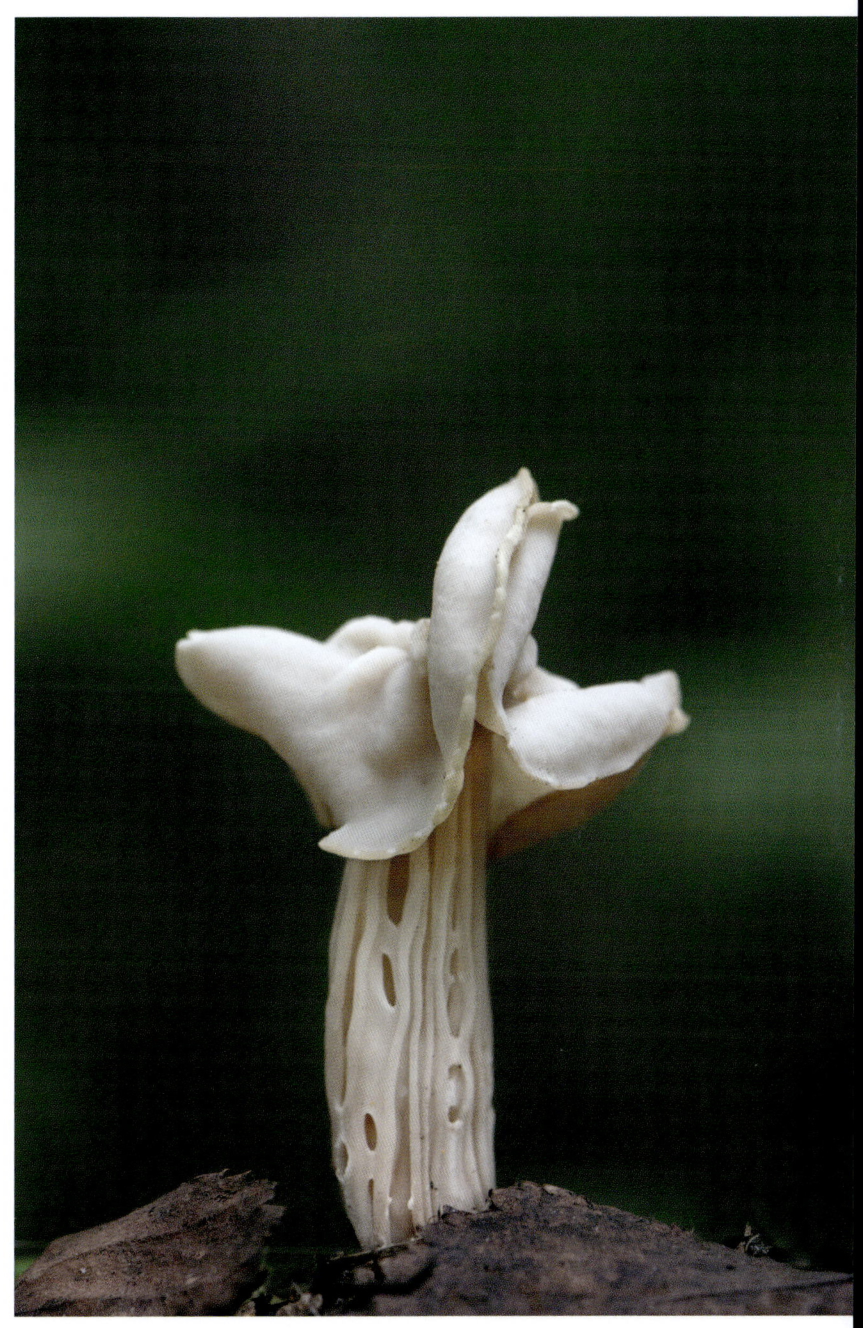

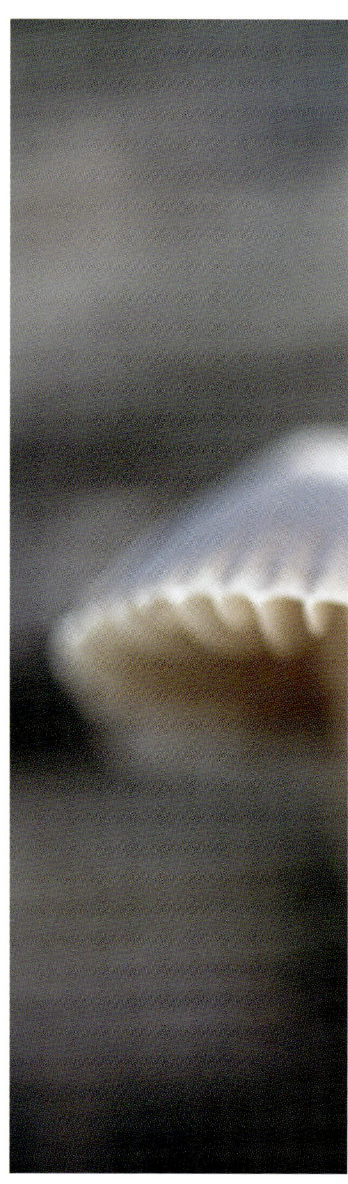

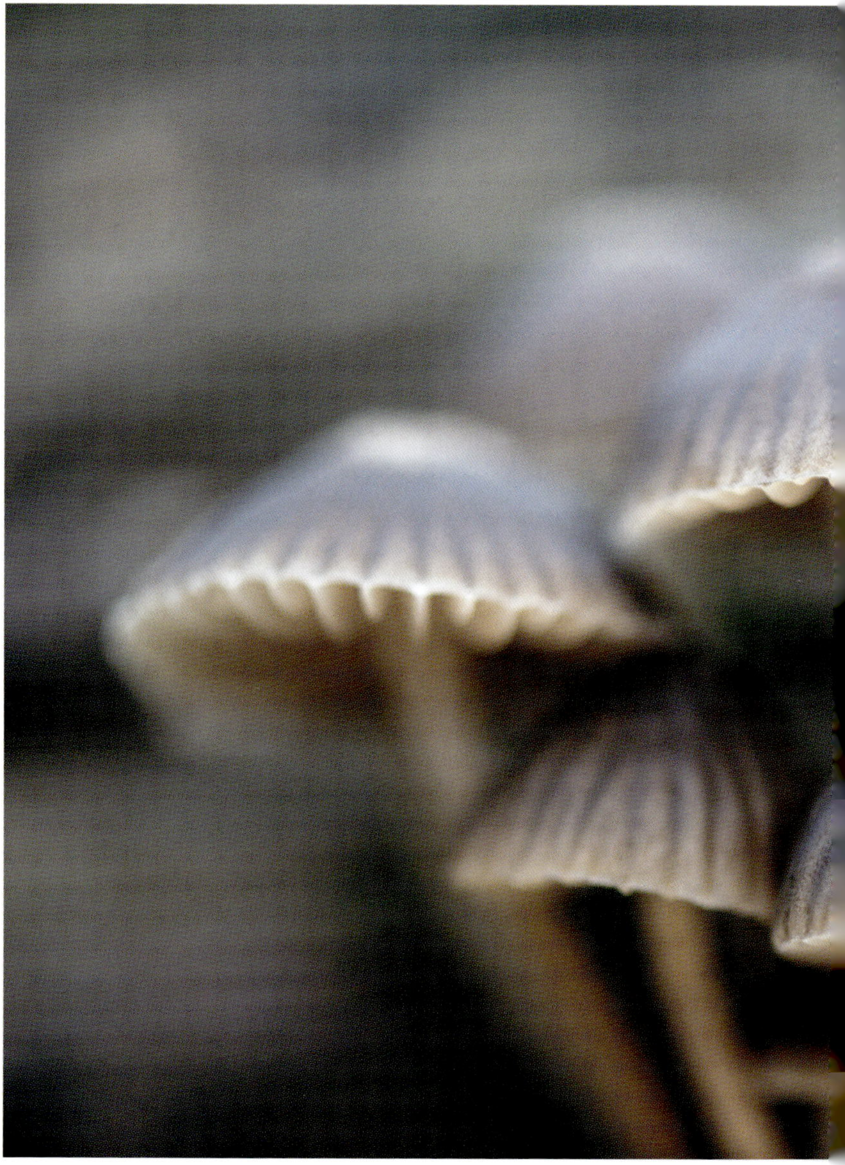

Umbrella-shaped mushrooms are an ingenious way to protect spores from getting wet. Bonnet gang (*Mycena subgalericulata*), Australia

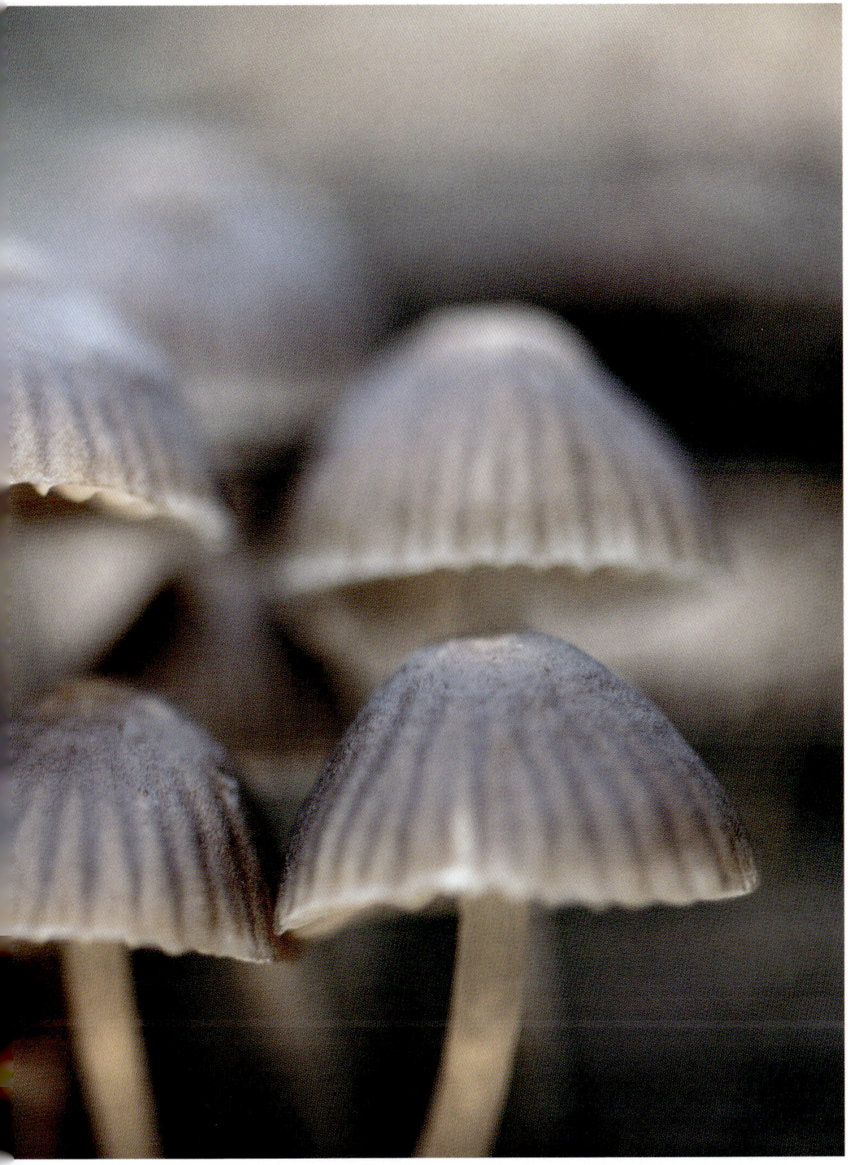

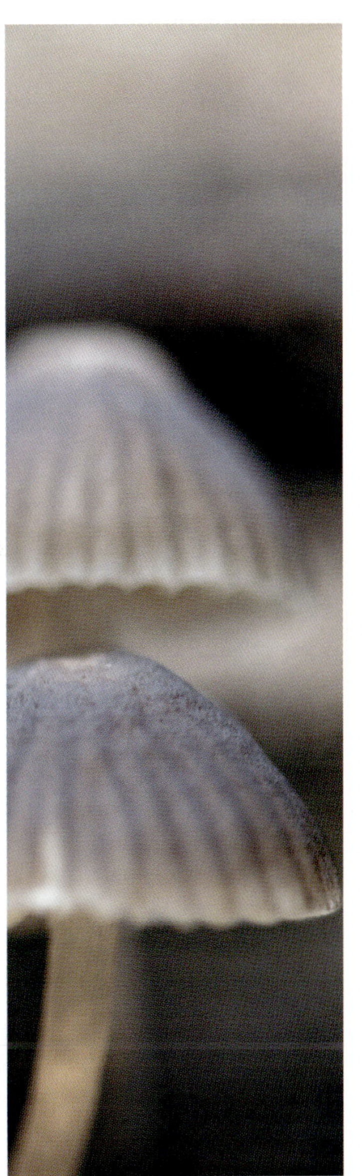

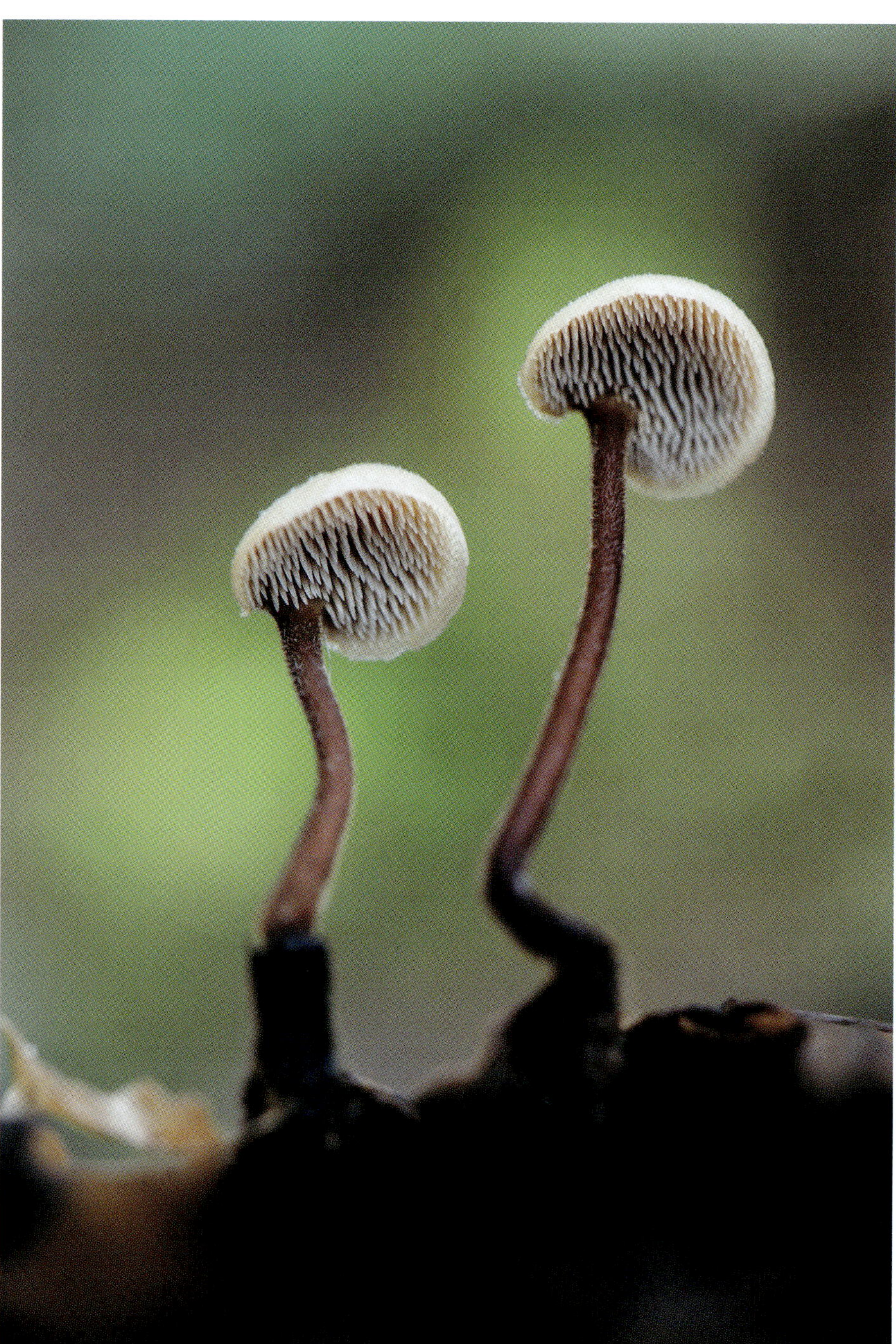

Cone tooth (*Auriscalpium vulgare*), USA (left)

Pixie's parasol (*Mycena interrupta*), Australia (opposite)

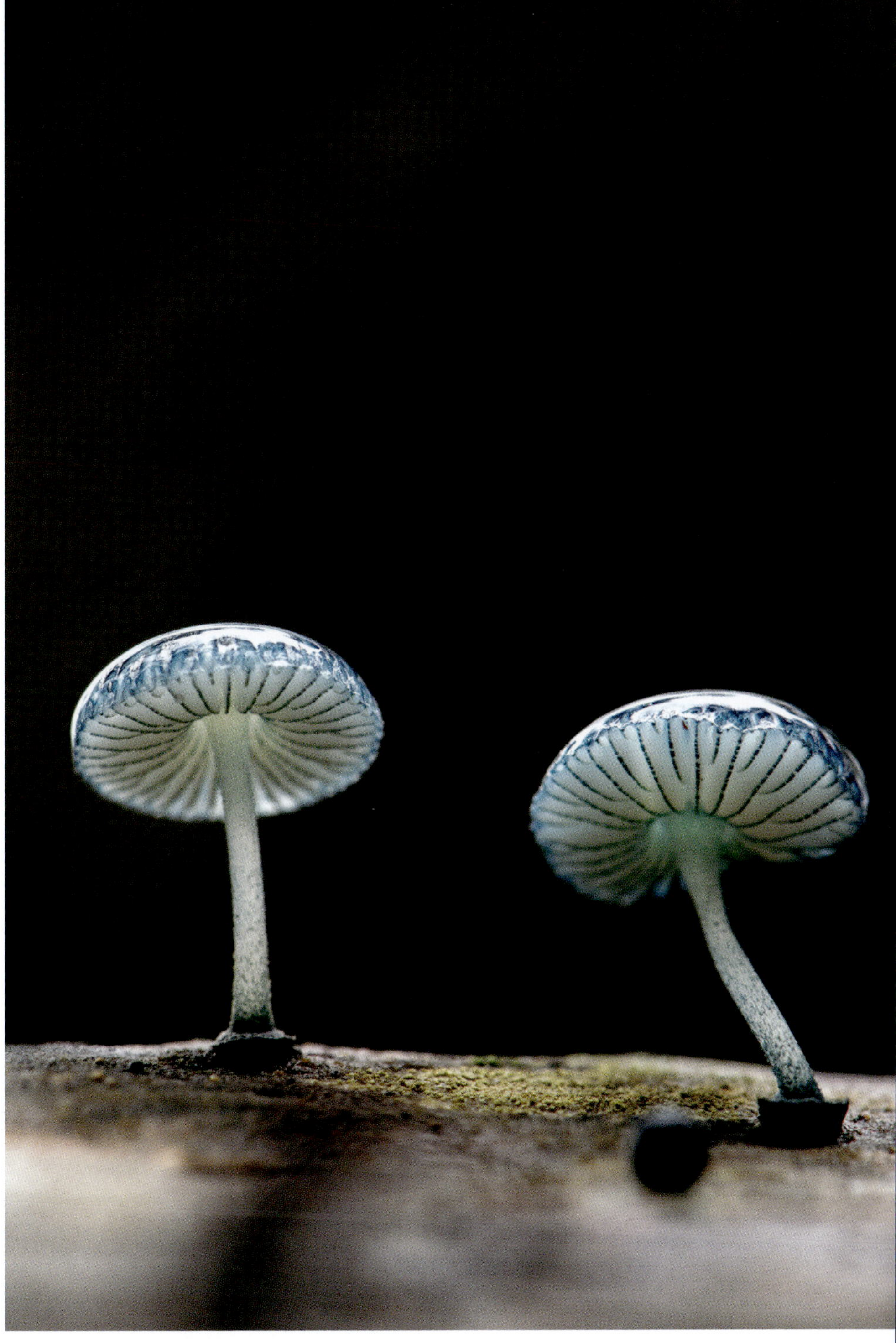

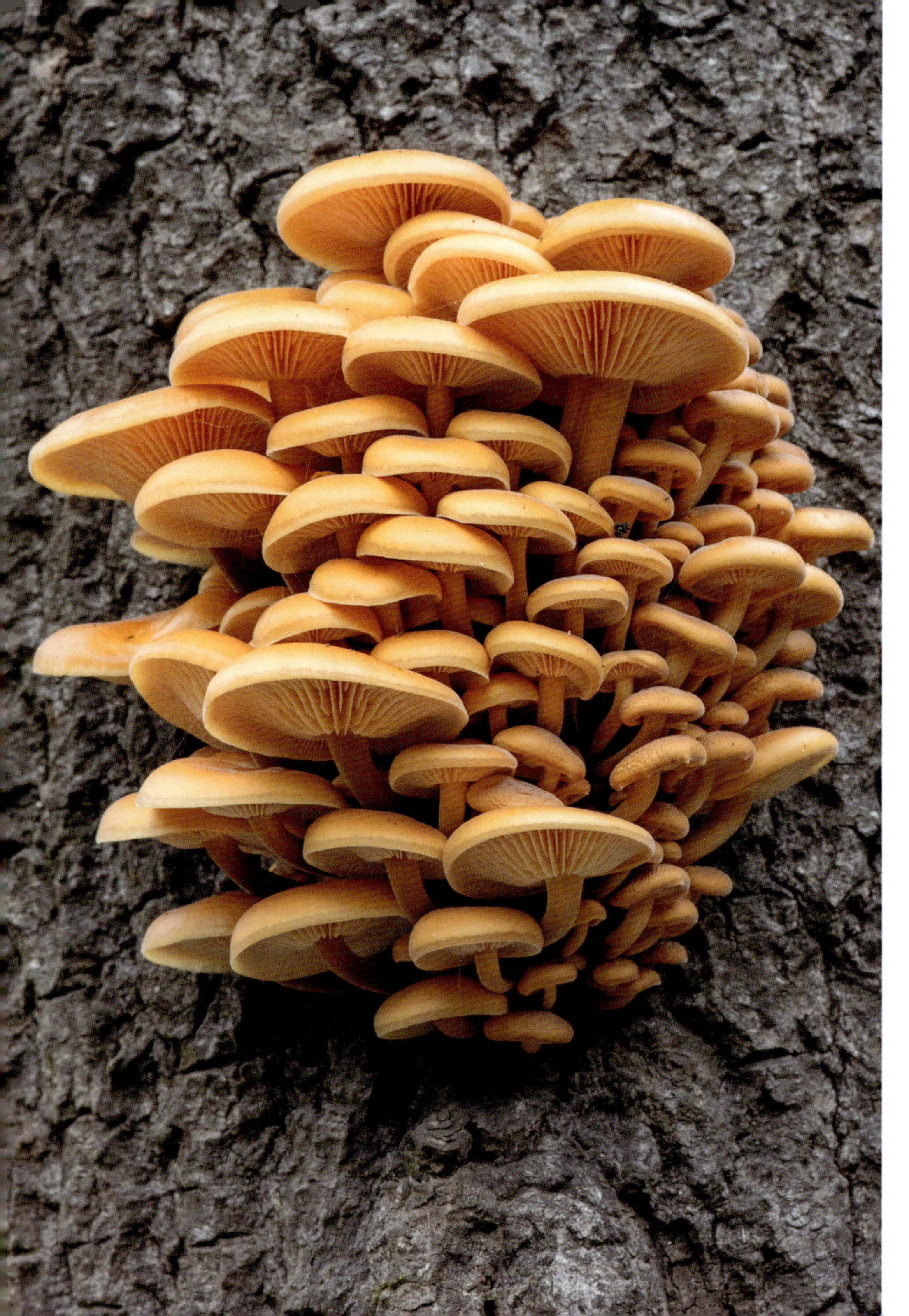

Porcelain fungus (*Oudemansiella mucida*), Switzerland (below)

Velvet shank (*Flammulina velutipes*), Italy (opposite)

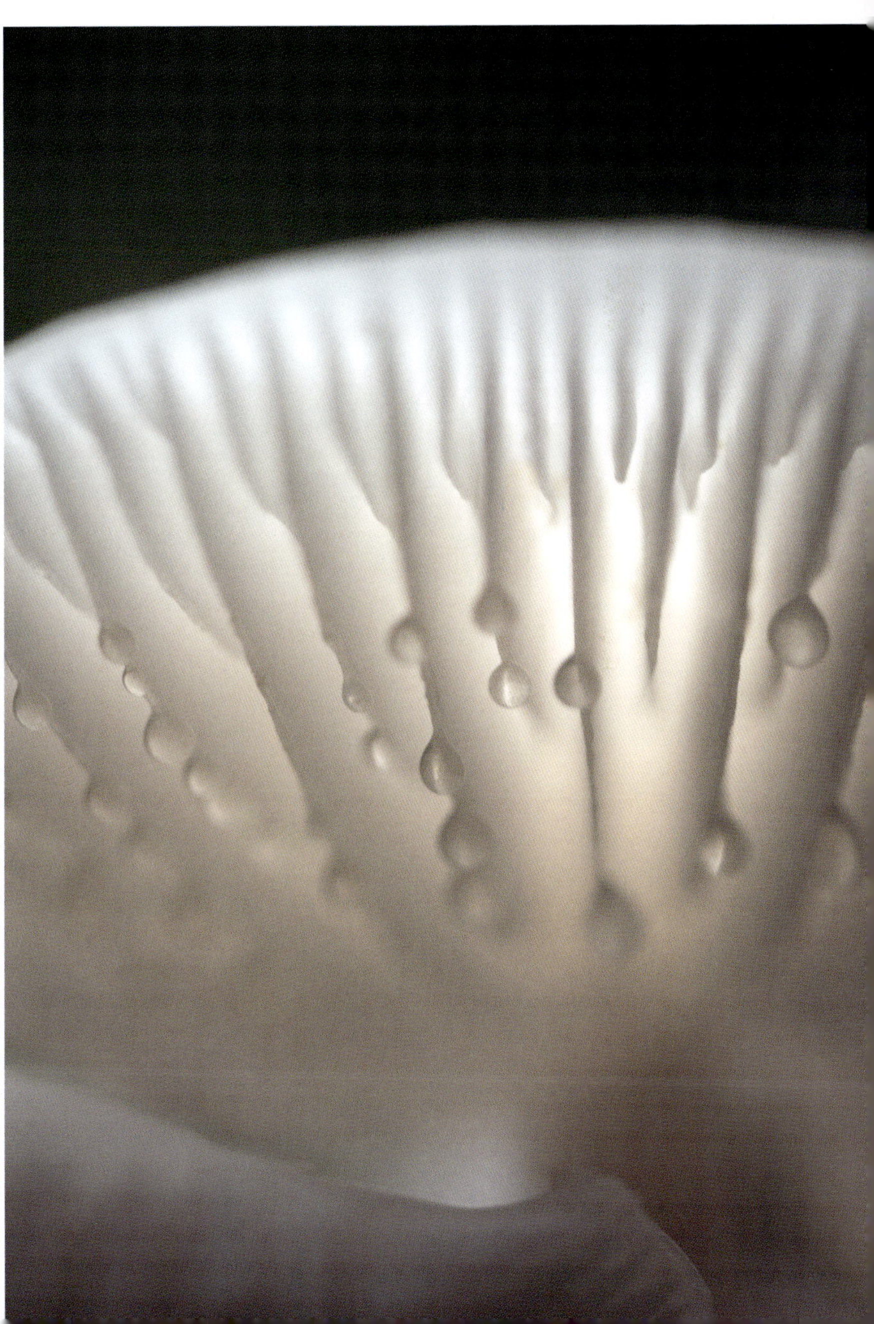

Shaggy parasol (*Chlorophyllum brunneum*) quartet, USA

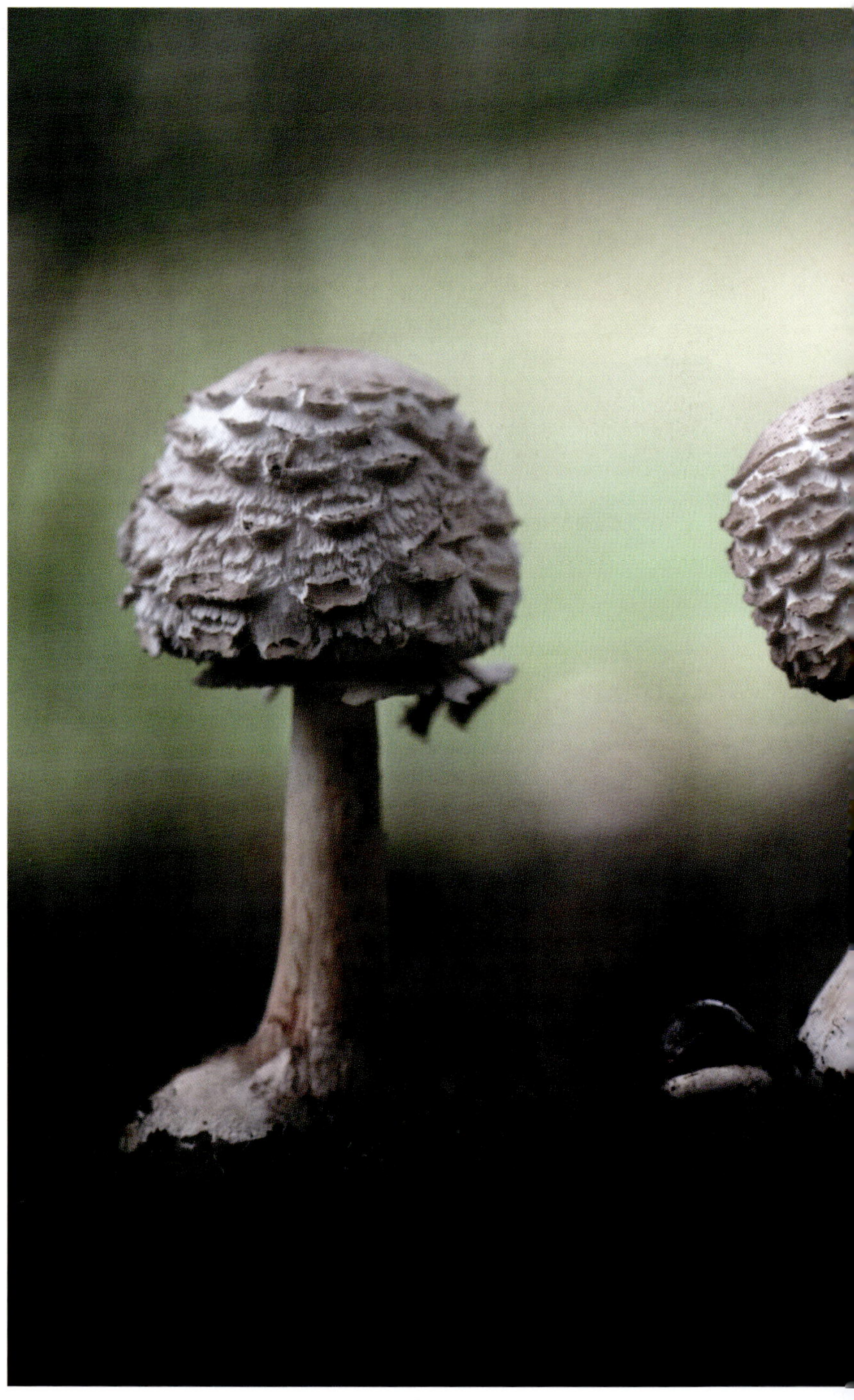

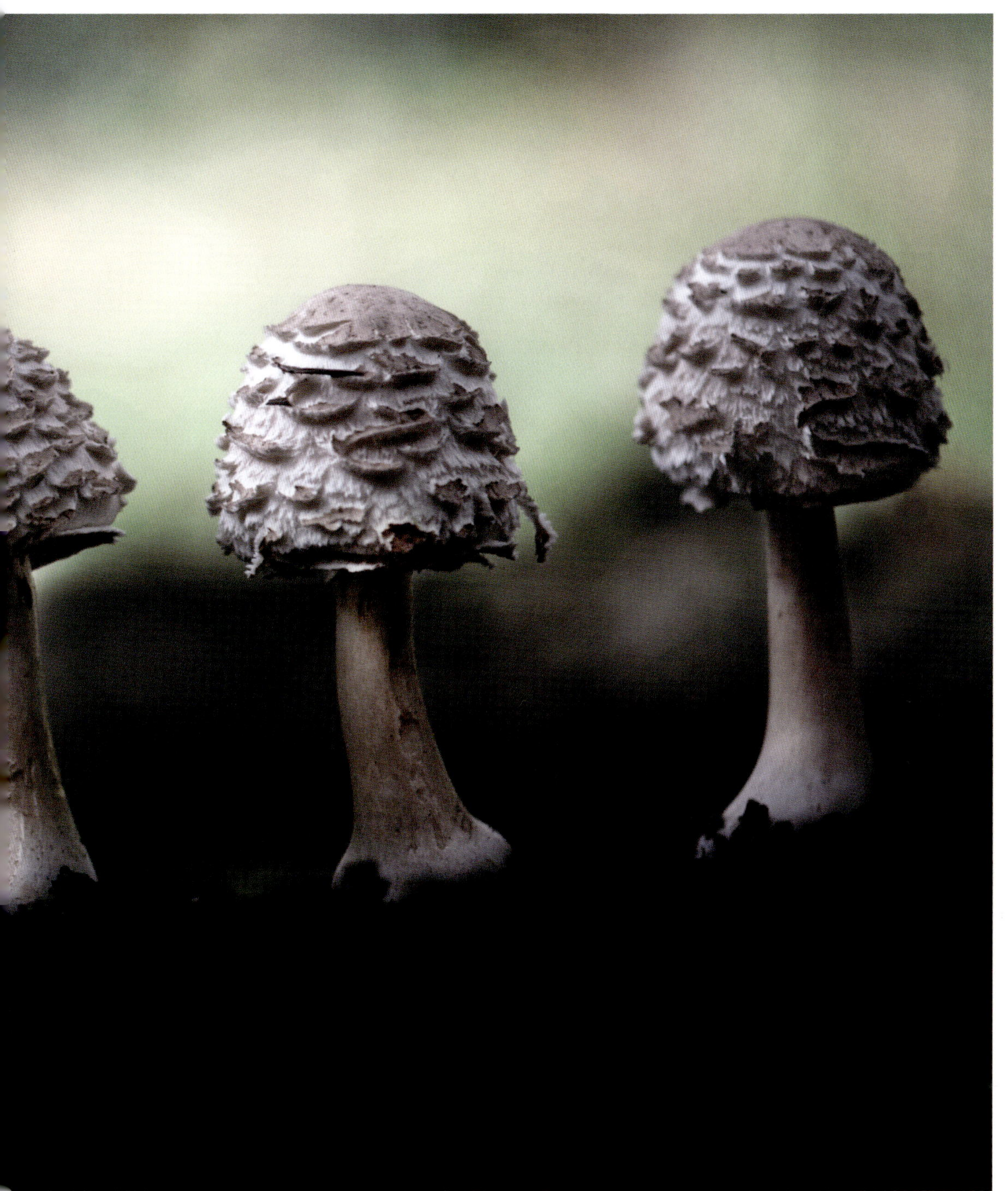

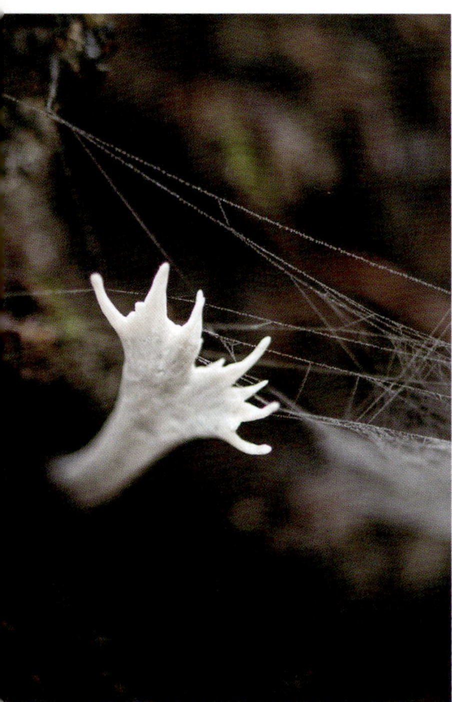

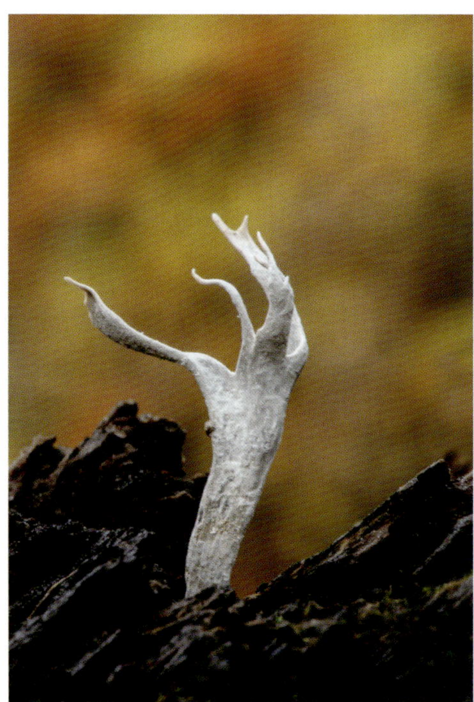

phenotypic plasticity

The candlesnuff fungus (*Xylaria hypoxylon*) can be highly variable in appearance

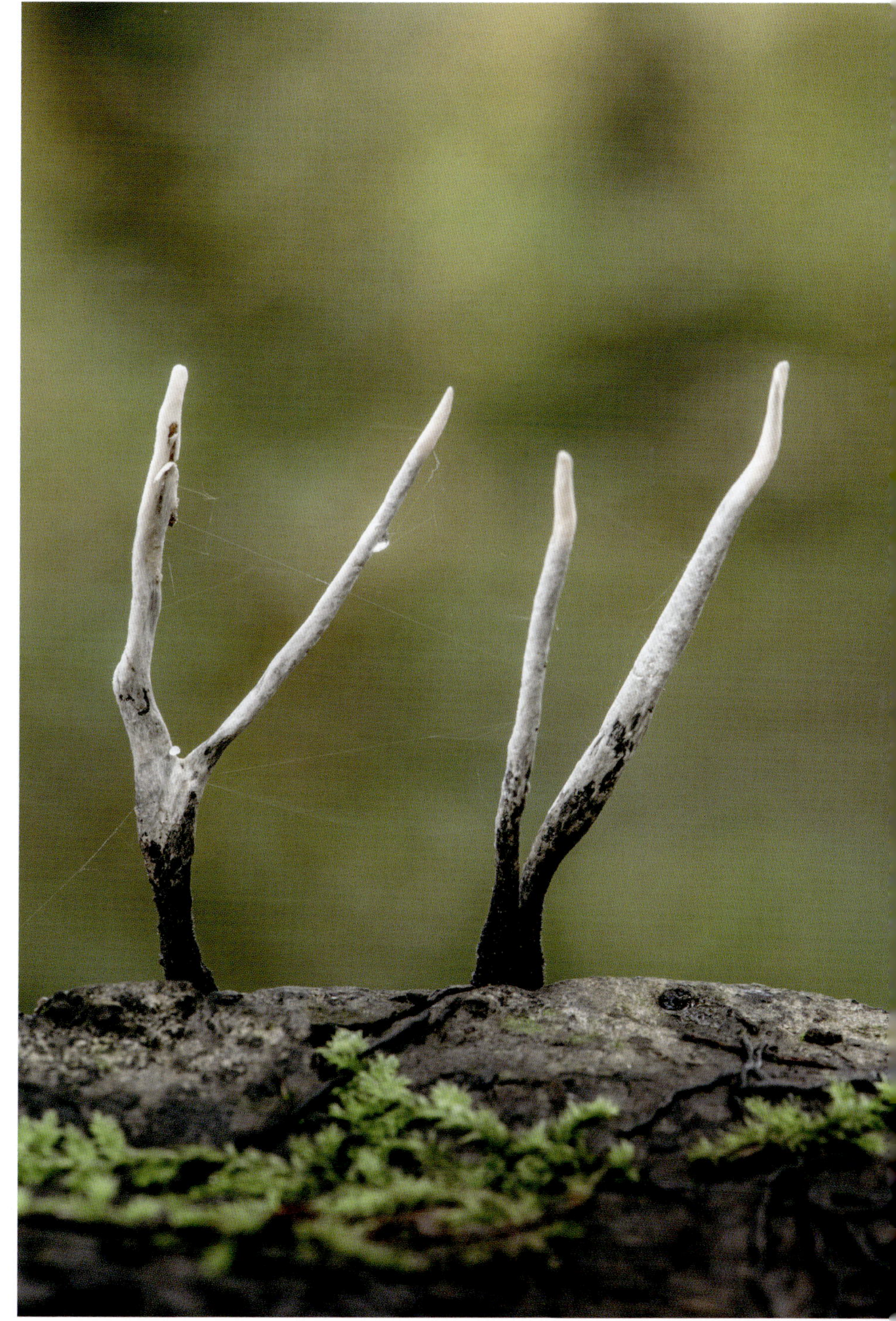

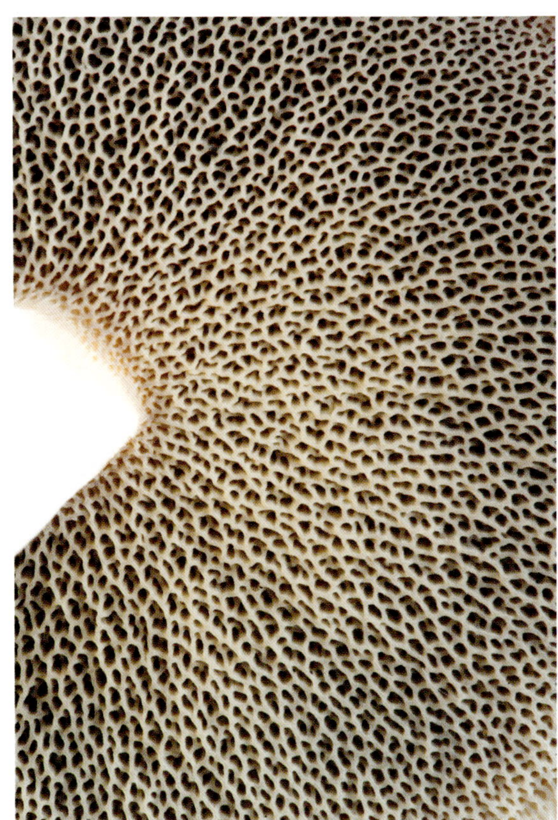

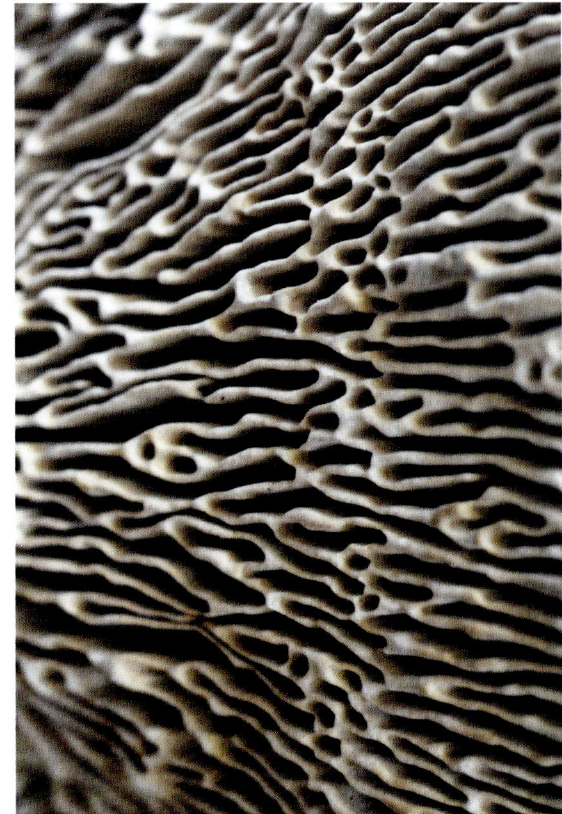

Left to right:

Peppery bolete
(*Chalciporus piperatus*), USA

Common mazegill
(*Daedalea quercina*), Sweden

Goat moth woodwax
(*Hygrophorus cossus*),
Switzerland

Sweetgrass tooth
(*Hydnellum suaveolens*), USA

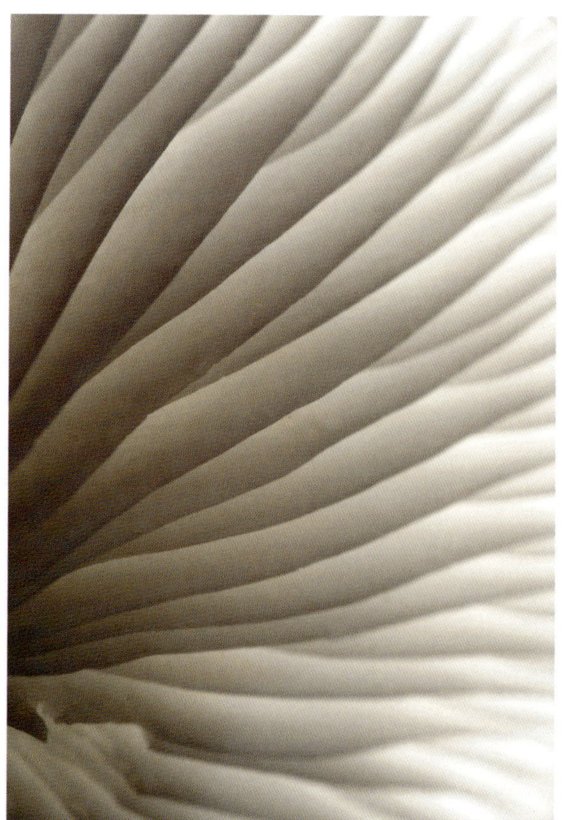

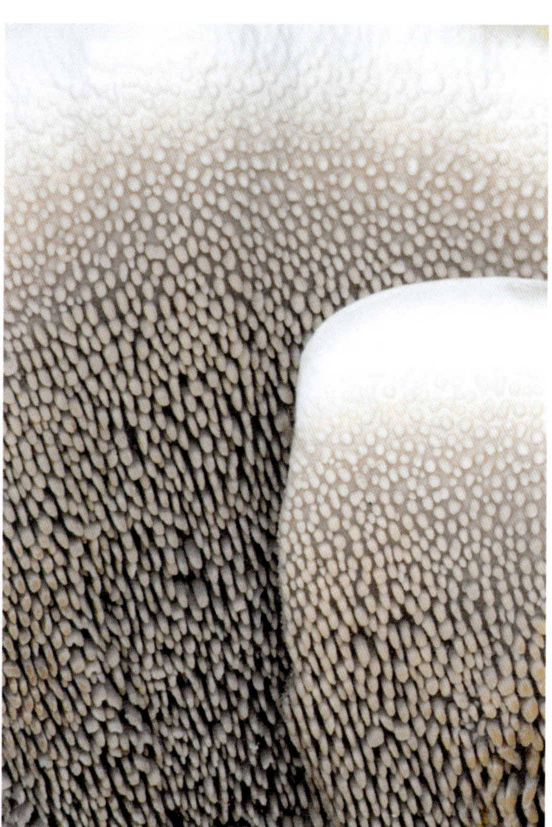

hymenia

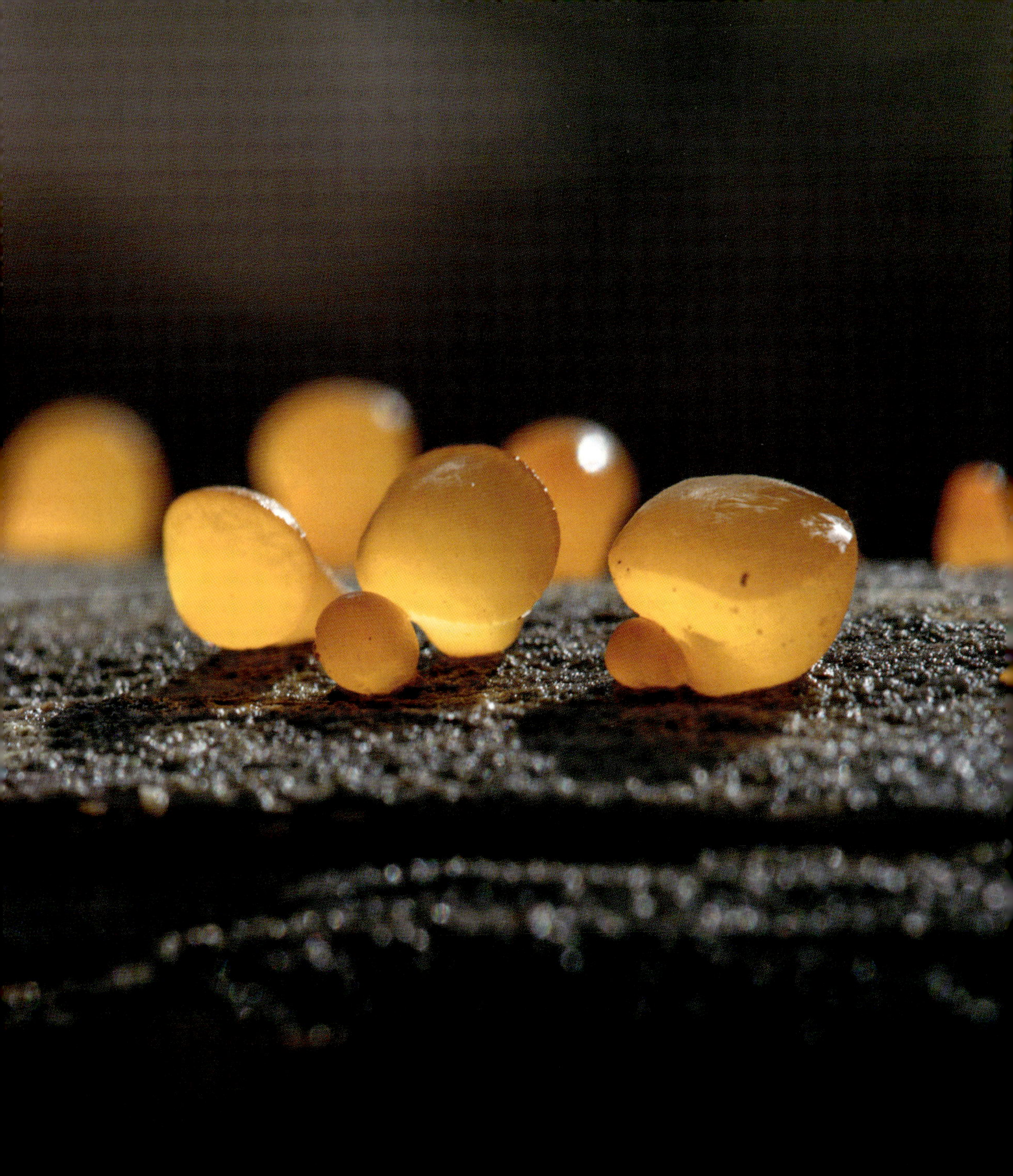

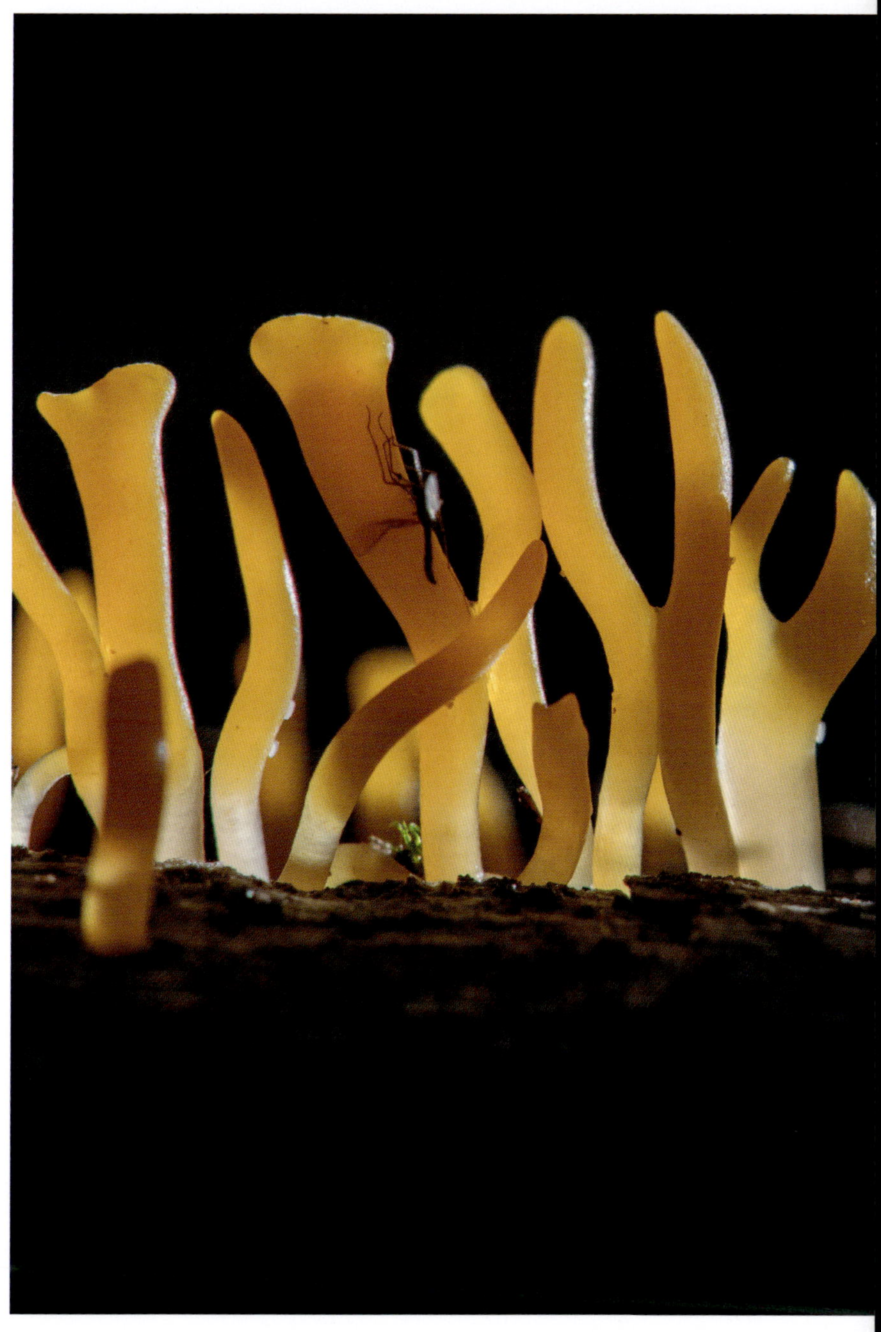

Pretty horn
(*Calocera sinensis*),
Australia (above)

Golden jelly bells
(*Heterotextus peziziformis*),
Australia (opposite)

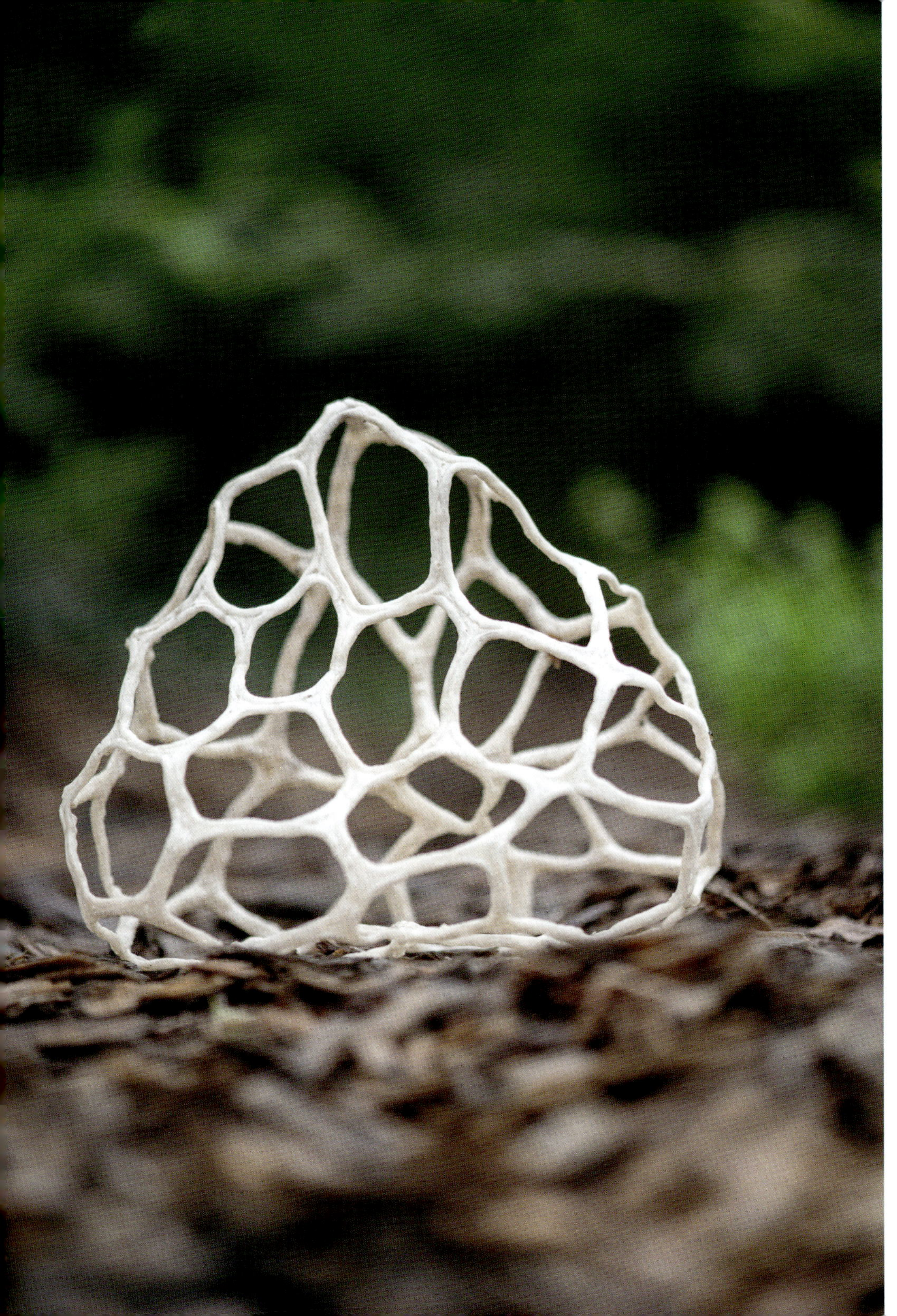

Smooth cage (*Ileodictylon gracile*), Australia (opposite)

Pipe club (*Macrotyphula fistulosa*), Sweden (right)

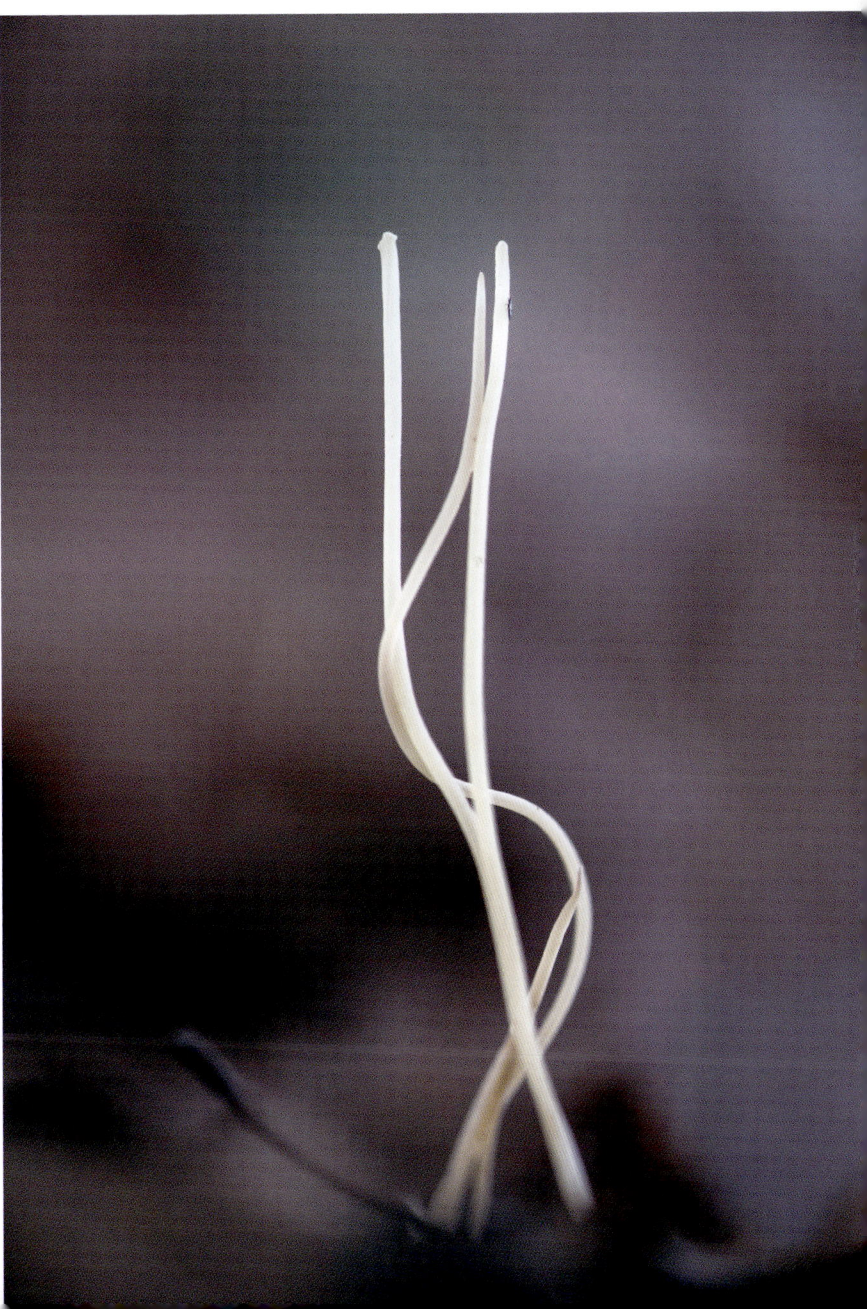

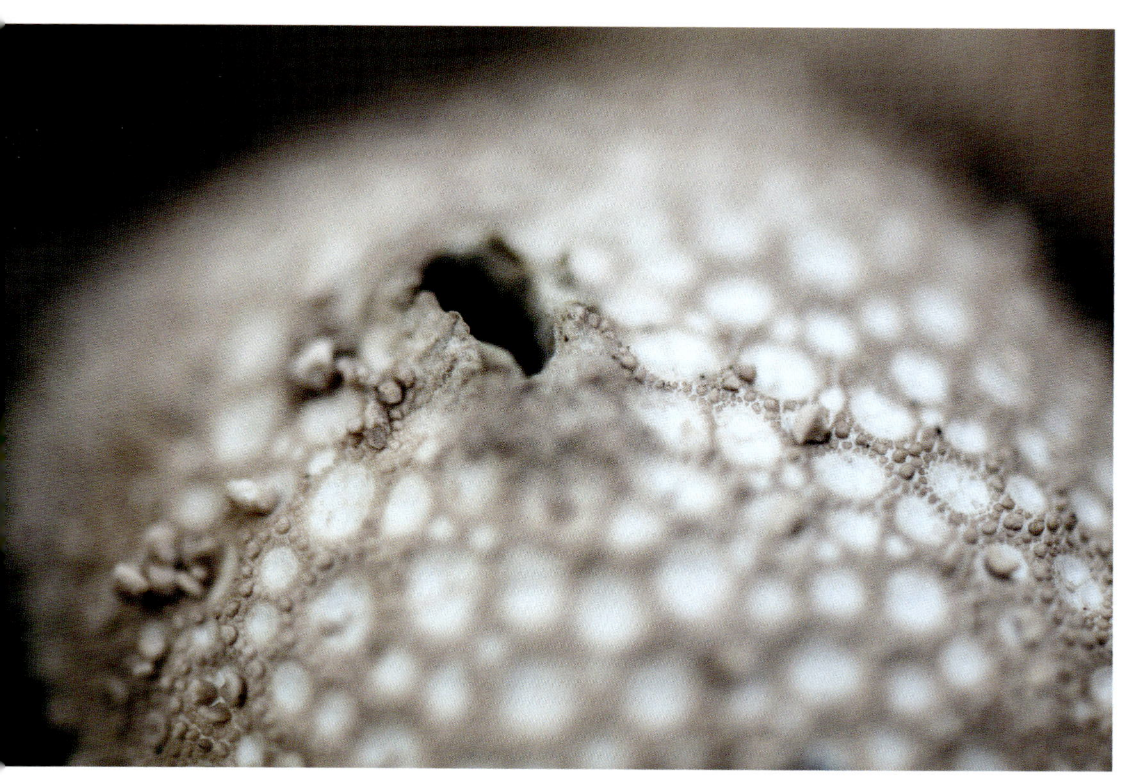

Gem-studded puffball (*Lycoperdon perlatum*), Norway

Following spread: Rainbow fungus (*Trametes versicolor*), Australia

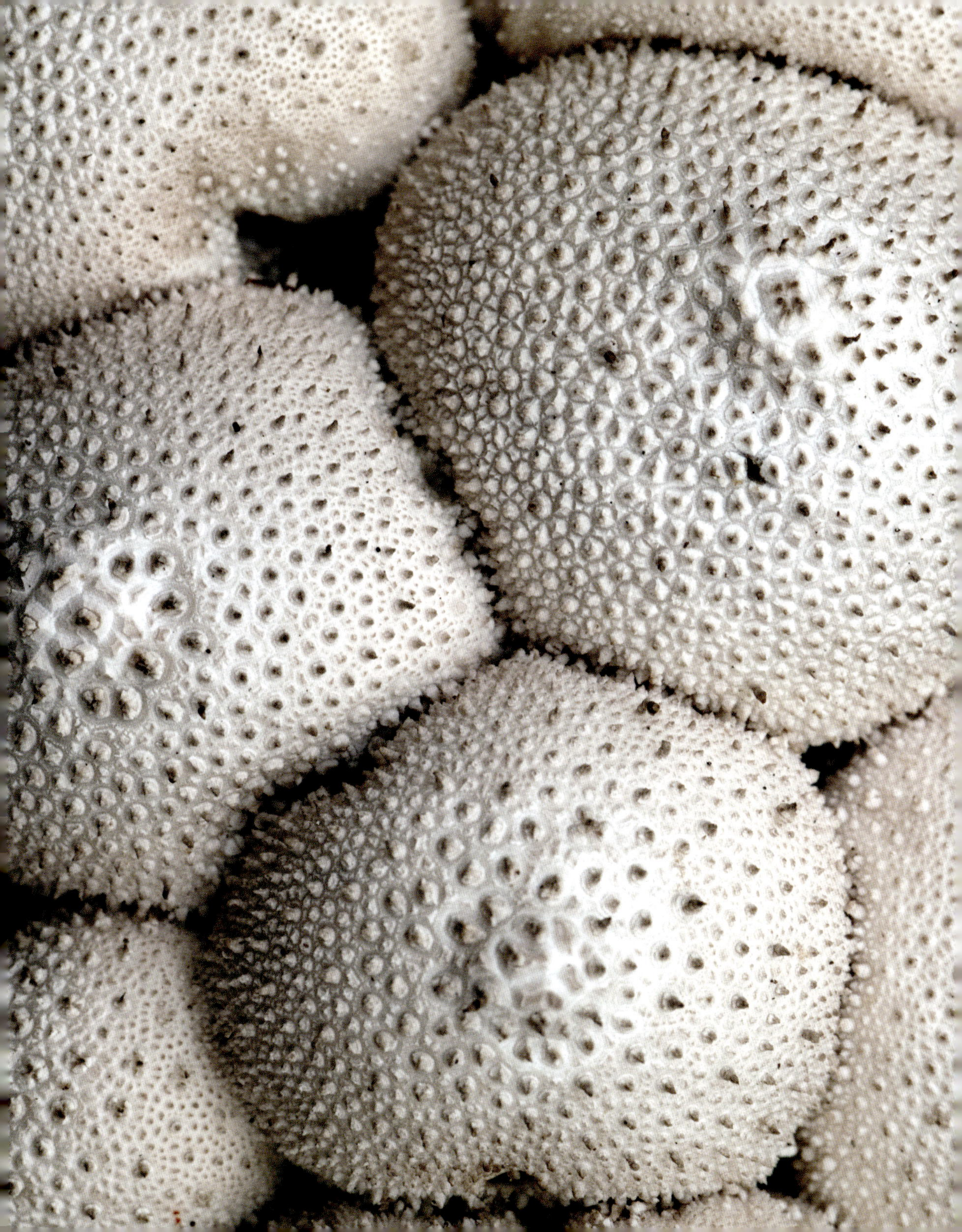

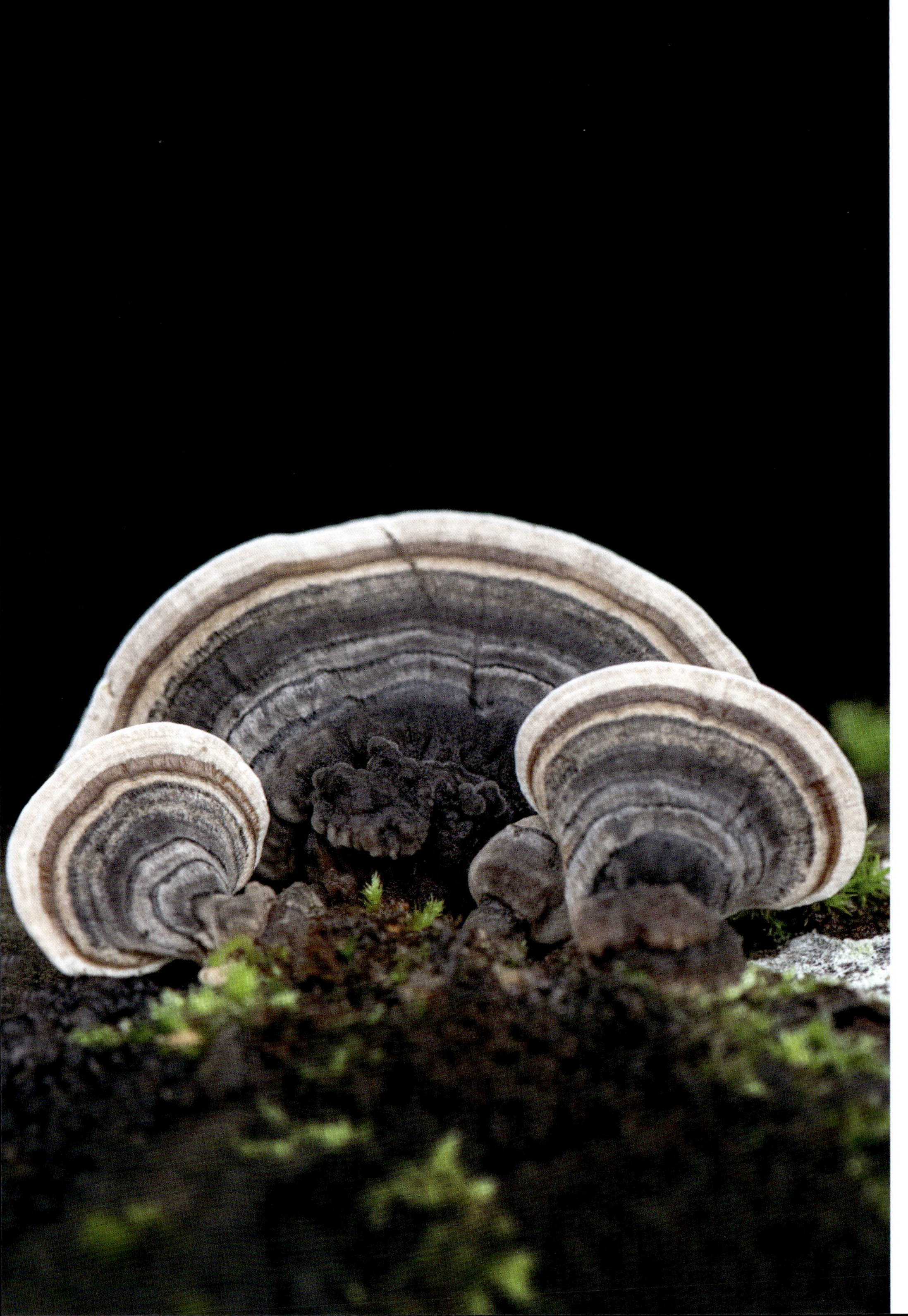

Spectrum

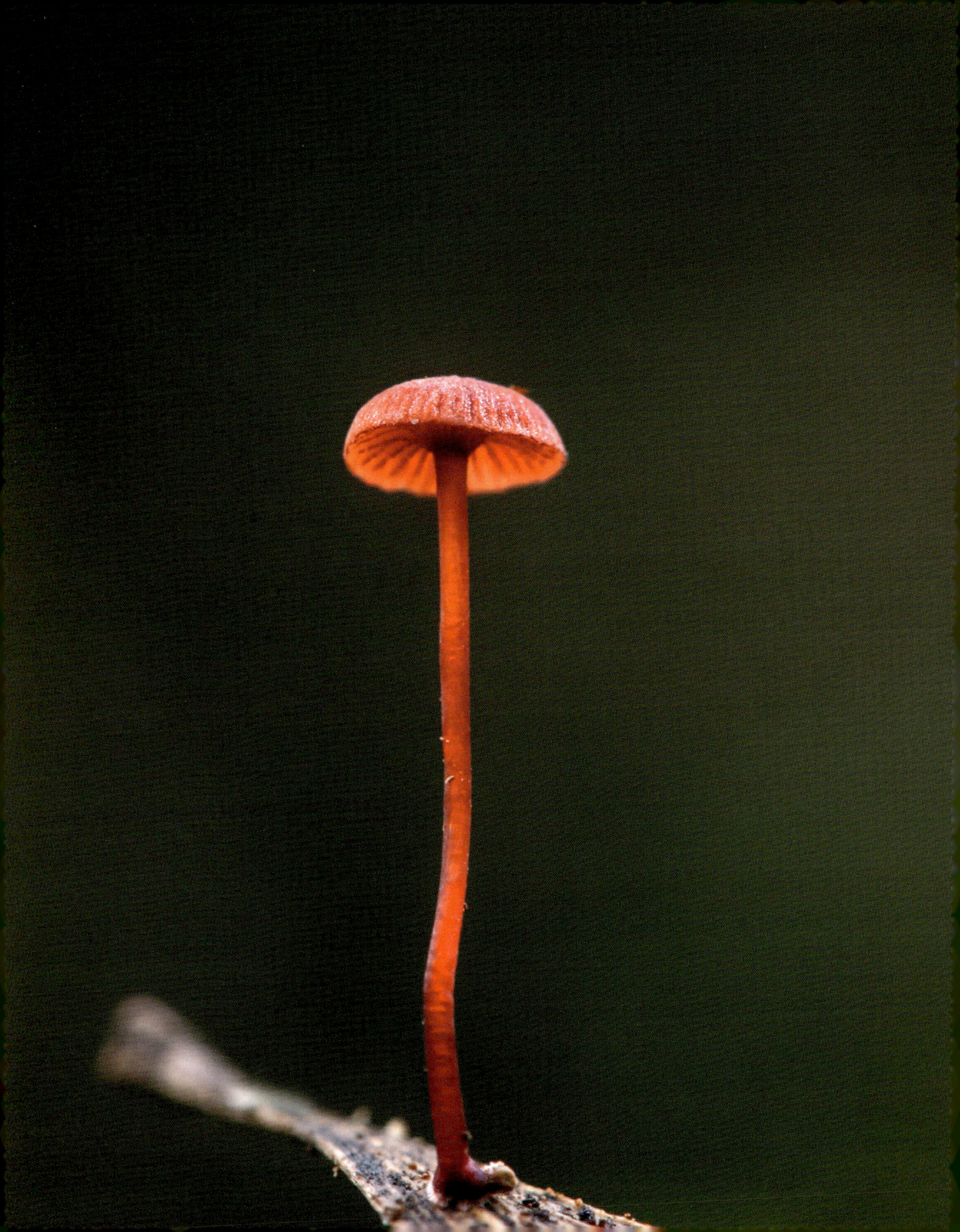

Ruby bonnet (*Cruentomycena viscidocruenta*), Australia

Life among *Homo sapiens*

I pushed aside the ivy to discover that someone had died in 1891. All I know is that their name began with TER; the remaining letters were obscured. A marble headstone, gilded in gold, marked their life, no doubt in the hope they would forever be remembered. As humans, we crave certainty and stability. We crave permanence. Many cultures commemorate the lives of lost loved ones with formalised rites and ceremonies, and the physical structures of graves and tombstones. And while we're alive, we invest vast amounts of energy, time and money to slow the decline of our minds, bodies and the objects we create.

Fungi don't always appreciate the human desire for permanence. As I wandered around Finland's Oulu Cemetery, lichens etched their own history into the gravestones. It's a slow process of transformation. Work in progress. Complexity increasing with growing disorder. As one creation comes undone, possibilities for new configurations arise. There's not much competition in real estate for smooth vertical stone, and lichens have free rein. You'd think they had expensive taste, given they favoured the incised gold lettering over the marble, but perhaps they simply grip the varied relief better than the smooth, slippery surface.

Nearby, a cemetery worker hacked away at unwelcome greenery with a noisy machine, and an old man pruned a rambling rose from a gravestone. Both attempted to suppress nature from obscuring or defiling their memorials. It's human nature to control, tend and care. But seen through another lens, the lichens create an enigmatic charm as the gravestones slowly metamorphose from lifeless entities to living ecosystems. Fungi and other agents of decay bring a particular aesthetic to human-made structures as they oxidise and corrode, disintegrate and dissolve. The headstones become hybrids: sculptures marking human history, evolving into living microhabitats. The ongoing succession of lichen colonisation and gradual remodelling remind us not just of the passing of time, but that transformation and flux are a normal part of natural cycles. I squatted down to photograph them and thought about the various lenses through which we might view them: each gravestone was at once a marker of multiple histories, a flourishing micro-community, a novel ecosystem, an indicator of environmental conditions and a hybrid work of art.

Whatever we manufacture, it seems there's a fungus that will have a shot at colonising it. Near the coastal hamlet of Argenton in France's westernmost region of Brittany, a speedboat lies high at low tide. An assembly of lichens occupies its plastic and paint, rubber and metal. The once smooth synthetics of this unexpected canvas are now mottled and textured with life – green and grey, ebony and orange. Seventeen thousand kilometres away in eastern Australia, lichens etch away abandoned cars in the bush. They create stunning tapestries, tiny microcosms of life among the rust and chips of paint.

Of all the fungi, it's mostly lichens that colonise our fabrications. But others do too. As we produce more and more synthetic compounds, and plastics accumulate in soil and water, scientists and bioengineers scramble for solutions to minimise environmental harm. Researchers at the University of Sydney found two fungus strains with powerful enzymes that can biodegrade plastics, including polypropylene, a plastic used in packaging and automotive parts that's difficult to recycle. Polypropylene accounts for 16 per cent of the entire plastics industry. This laboratory experiment produced exciting findings, especially given the sobering claim that the millions of tonnes of plastics in the oceans could soon surpass the mass of fish.

Yet we still don't understand how fungi break down plastics in the natural environment. While the technology is promising, scaling up production on a commercial scale, and overcoming political, social and behavioural responses to plastic consumption present further challenges. There is, of course, an undeniable appeal in finding a fungal solution to the scourge of polluting plastics. But is a technological 'fungal fix' to our voracious consumption of plastic the best approach, given that the pace of plastic production far exceeds that at which fungi can dismantle them? Stopping or reducing production in the first place and possibly replacing plastics with mycelial alternatives seems like the better options than an end-of-production 'tailpipe solution'.

Upheaval

In early autumn 2019, I stood on the dry margins of Alder Lake in the US state of Washington. Before me lay amputated alder stumps, strewn like dismembered limbs. The trees had drowned when the lake was enlarged during the construction of the Alder Dam in 1944. Their roots, once entwined with mycorrhizal fungi and nitrogen-fixing bacterial partners, lay lifeless, twisted and exposed. I poked among them in search of fungi but found none.

Human disturbances that damage or destroy habitats radically change fungal communities. Depending on the degree, extent and intensity of disturbance, many fungi are eliminated. Some are more resilient and will recolonise because of the ubiquity and resilience of spores that blow in or are transported by vectors from elsewhere. But there's a limit to how long this can be sustained if the places they come from are also repeatedly disturbed by agriculture, forestry or urban developments that smother spores and mycelium beneath concrete. Although we make worthy attempts to restore damaged environments, rates of destruction exceed rates of restoration.

Across hemispheres, in Australia, more than 200 years of trampling by hard-hooved stock has left irreparable damage to soils and the fungi that inhabit them. As soil compacts, it squashes out air, prevents water infiltration and leads to the diminishing of complex ecologies as soil reverts to dirt. Seedlings fail to regenerate, leaving the soil surface bare. Erosion sets in. Some fungi can recover if the pressures are eased, but many cannot.

From Red Rock Lookout in south-western Victoria, crater lakes are visible among the scattered volcanic peaks. Nine lakes within the volcanic plains region are listed under the Ramsar Convention – an international treaty for the conservation and sustainable use of wetlands – and they host an array of waterbirds at critical life stages, such as migration and breeding. The wetlands provide an oasis among a sea of agriculture and altered landscapes. When I visited in 2007, the area had suffered 13 consecutive years of below-average rainfall. The previous year had been the driest on record. Most of southern Australia was affected by the 'millennium drought' (1997–2009), and some farmers told me it was not only the worst drought they'd ever experienced but the worst since European settlement. From the lookout, the landscape stretched before me in tones of beige and brown. There was barely a blade of grass to be seen. The tracks of hungry sheep scarred the lifeless paddocks. From the opposite bank of a dry and eroded riverbank, the desperate creatures stared at me as though I could offer them some sustenance. While several El Niño weather patterns – resulting in hot, dry conditions in eastern Australia – occurred during the drought, meteorologists attribute its extent and severity to climate change.

Eventually, the rains returned, the paddocks greened and the lakes filled again. Australia's biota and landscapes are adapted to the vagaries of weather and climate. Ecosystems recover. But when they lack the infrastructure in soils and the functions provided by fungi, the process is slower and rarely complete. Restoring fungi and other biology in soils builds resilience and speeds recovery.

Bringing back fungi

People often ask me, 'Where can I find fungi?' The question always seems a little odd, as there are few places where fungi don't grow. However, the more we disturb and damage soils, and the more of their food we steal – leaf litter and logs and everything in between – the fewer fungi you'll find. Fortunately, a shift in thinking sees some gardeners and farmers, conservationists and land restorers, encouraging fungi in damaged landscapes to kickstart recovery. Most terrestrial, environmental restoration focuses on revegetating, but for plants to thrive, it's best if foundations provided by fungi, invertebrates and their kin are first put in place. Simple changes to the ways we restore damaged ecosystems and manage gardens could have dramatic effects. Opting for ecology over aesthetics maximises the chance of recovery. Providing diverse habitats and microclimates, and resisting the temptation to woodchip, tidy and manicure, encourages conditions for colonisation by diverse fungi. It begins with aligning our environmental aesthetic to mimic the dynamism and messiness of natural systems. While we can help to create novel habitats, we might then step back and allow them to maintain themselves. It's tricky to navigate the tension between attentive care and actively letting go, but in doing so we might discover unexpected affinities with fungi as a vibrant interface between disturbance and recovery.

My work across hemispheres and landscapes with scientists, forayers, foragers, First Nations people and others helps me attune to how fungus distribution plays out in time and space. Photographing fungi exposes temporal extremes – from the fraction of a second of a shutter speed to the giddying geological scale of TER's lichen-colonised gravestone. Fungi subvert our efforts to grasp and calibrate time. Although mushrooms appear only fleetingly, the potentially perpetual existence of a fungus mycelium upends ideas about lifespans. So long as there is food available, technically a mycelium can live forever – but only if it's allowed to do so undisturbed.

It has taken a long time to shift thinking from fungus as disease or destroyer to fungus as vital organism underpinning terrestrial ecosystems. While much of the current environmental narrative is a crisis narrative that speaks of loss, what if we reframed the largely negative lens of fungal decay to one of possibility and creative transformation? What if we reconceptualise the idea of 'fungus as tree-rotter' into 'fungus as tree-hollow creator'? Or 'lichen as grave obscurer' into 'lichen as ecosystem starter'? We could inflect the decomposing ability of fungi a little differently and see it as part of the constant recycling and renewal of life. Perhaps then we might spend less energy resisting natural processes and accept that order is only ever temporary and artificial. The vast scale and distant top-down view from which environmental issues are often visually portrayed can detach us physically and emotionally. But fungus photography can help bridge the scalar divide. Simply noticing and getting in close can make tiny lives perceptible, and can ground us.

Reimagining fungi challenges us to become comfortable with impermanence, to accept that life is transient. When helping fungi and ecosystems re-establish, we might consider what we can contribute, while questioning whether ongoing intervention is always the best approach. Forests and other natural ecosystems work best when left alone and allowed to get on and grow old in their own time. Given we're relative newcomers on the planet, perhaps it's time to share the reins.

Following spread:
Lichen-colonised
gravestones, Finland

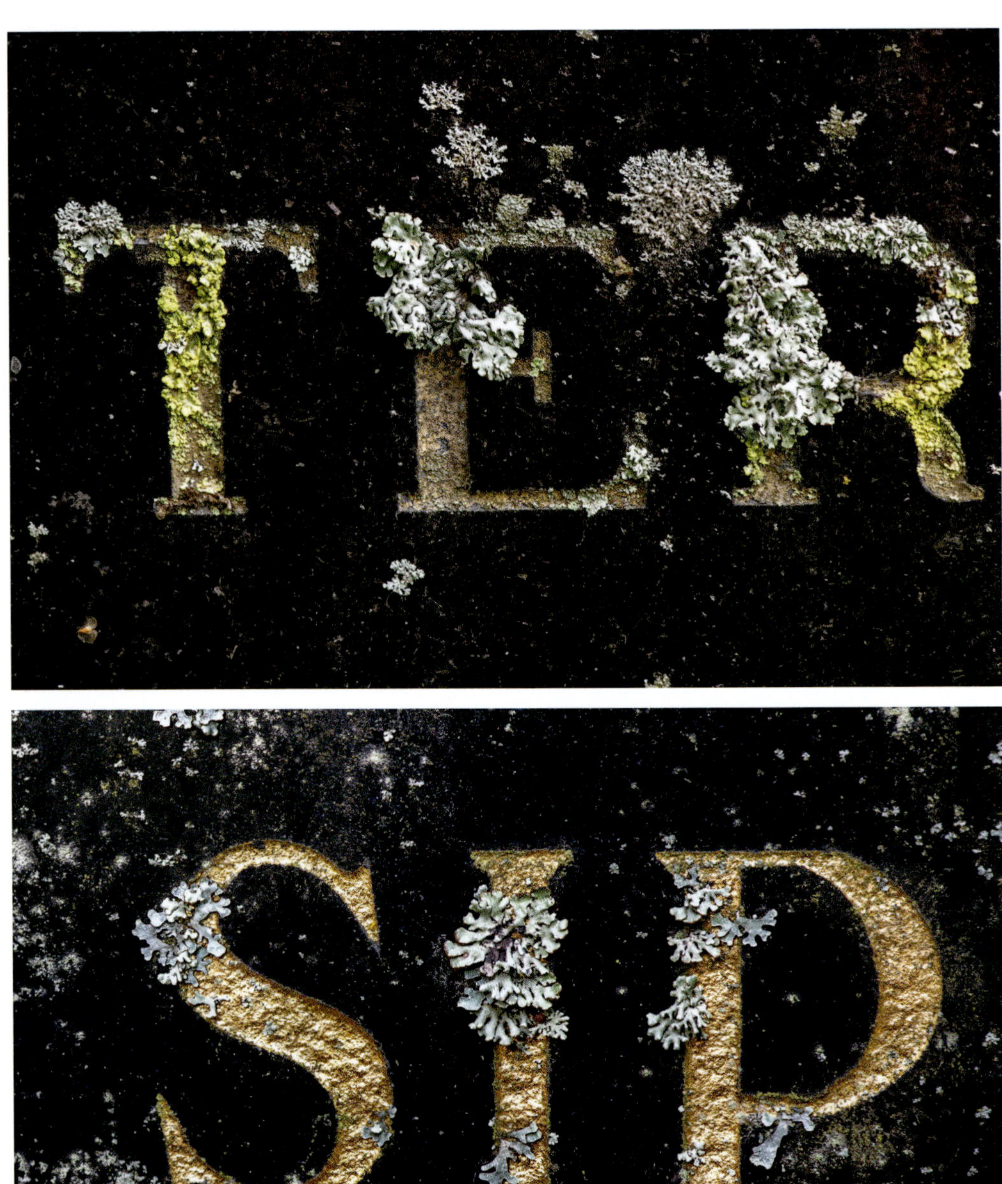
TER
SIP

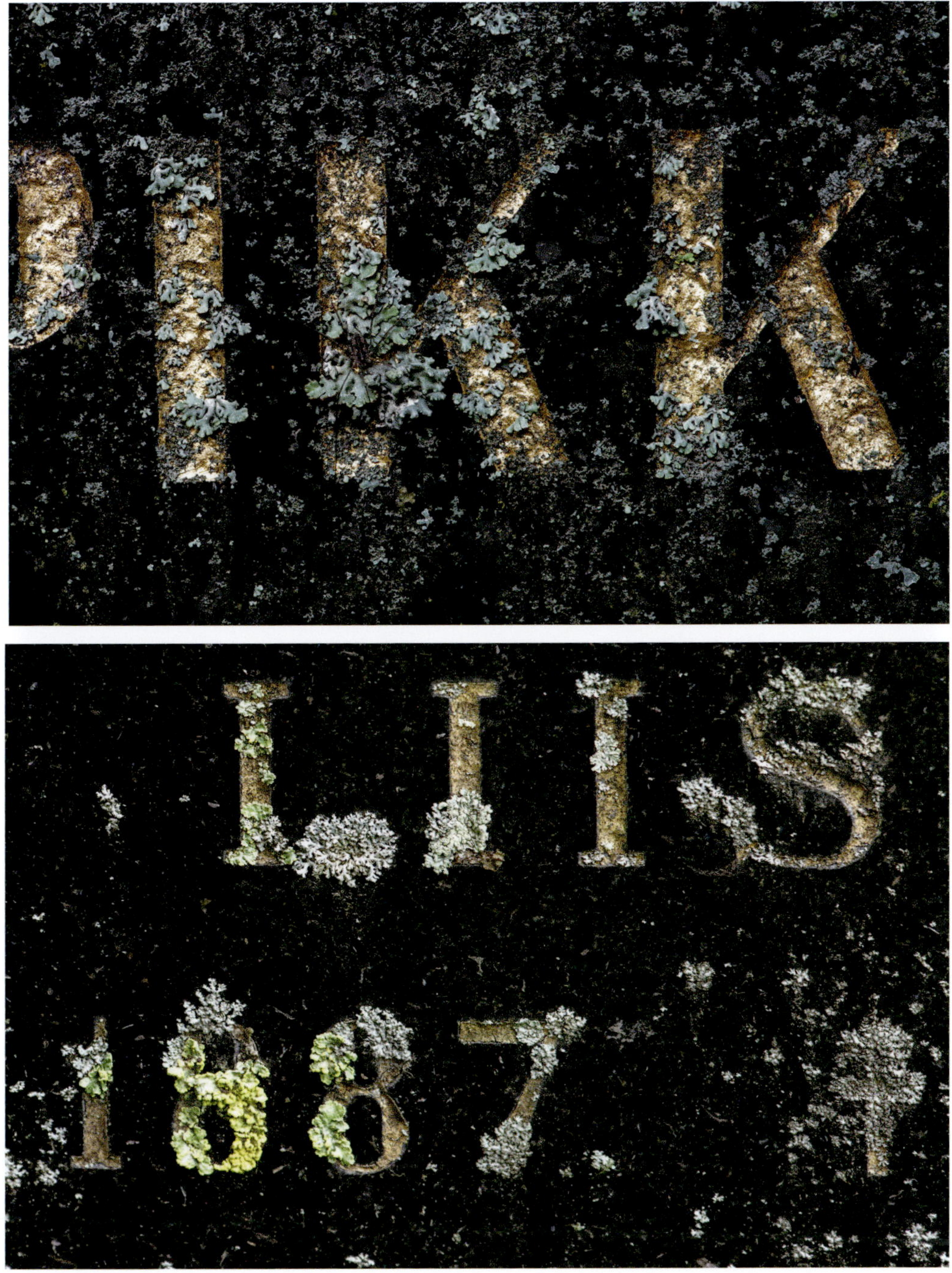
LIIS

Lichens colonise a speedboat, Argenton, France

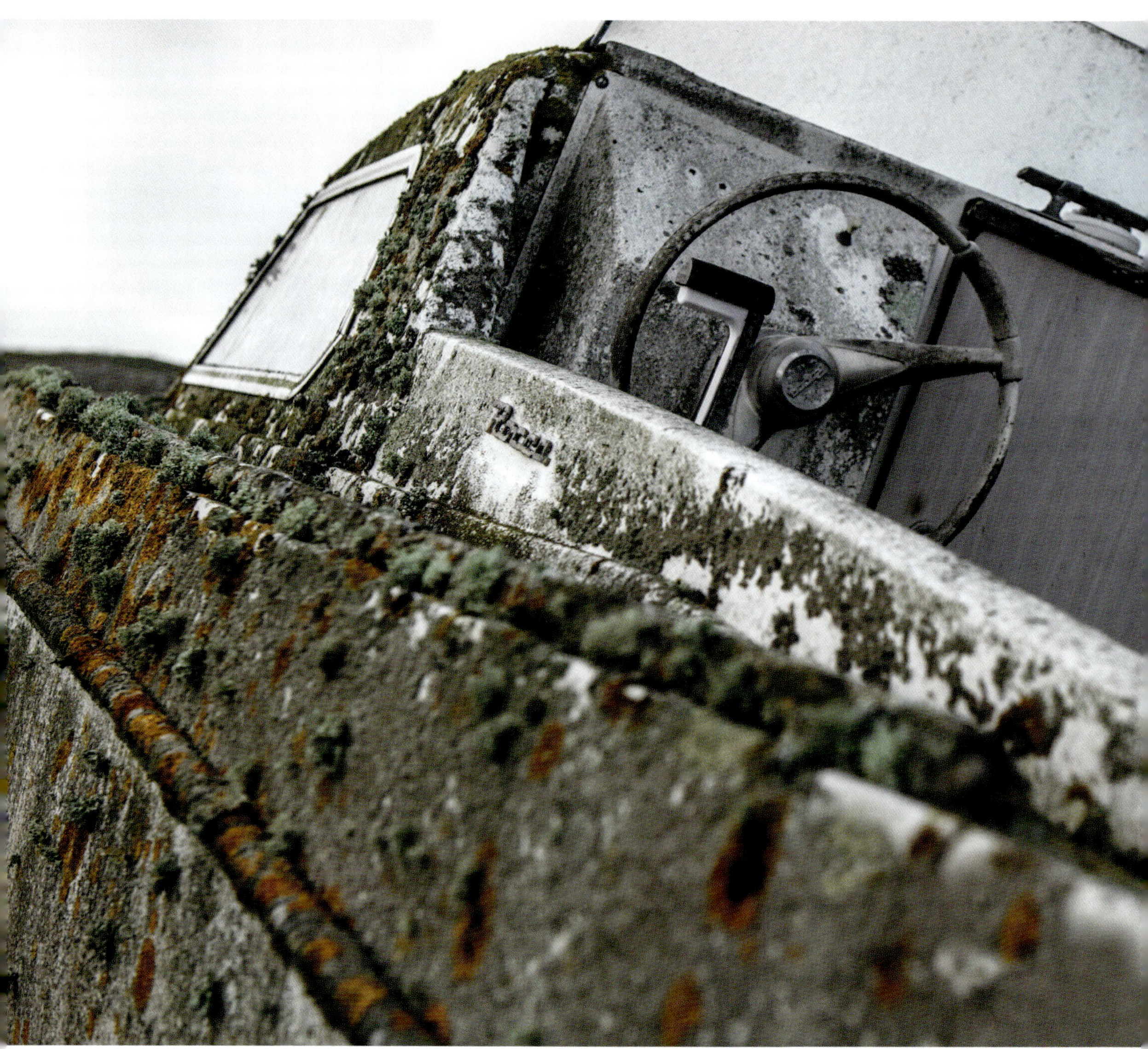

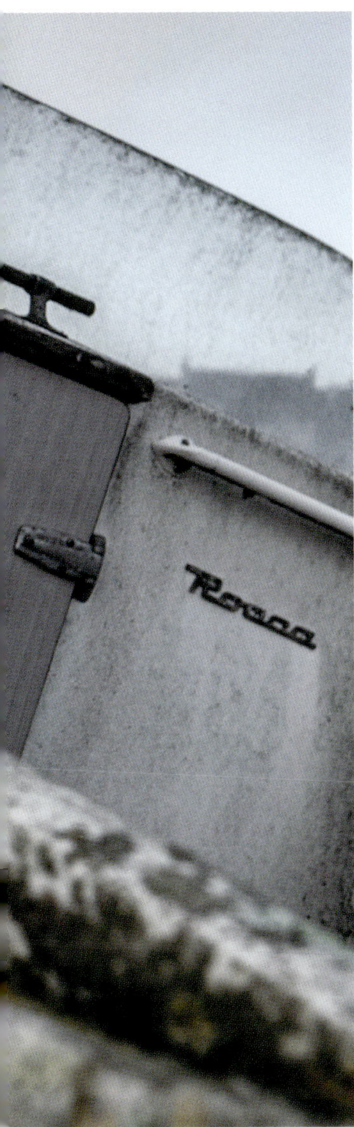

Following spread:
Lichens on old cars,
Australia

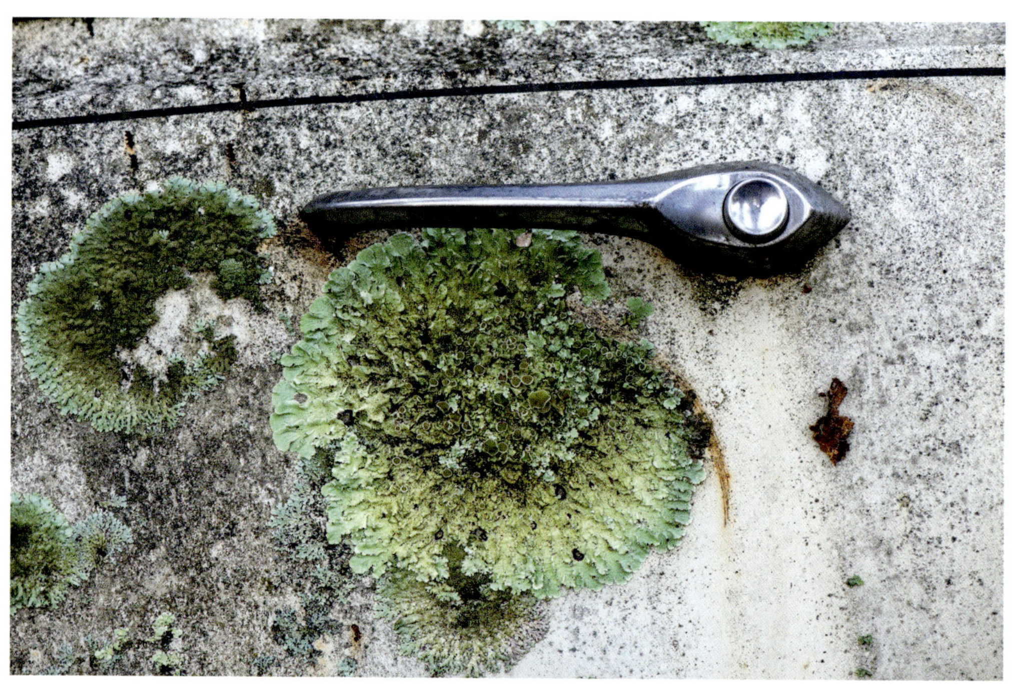

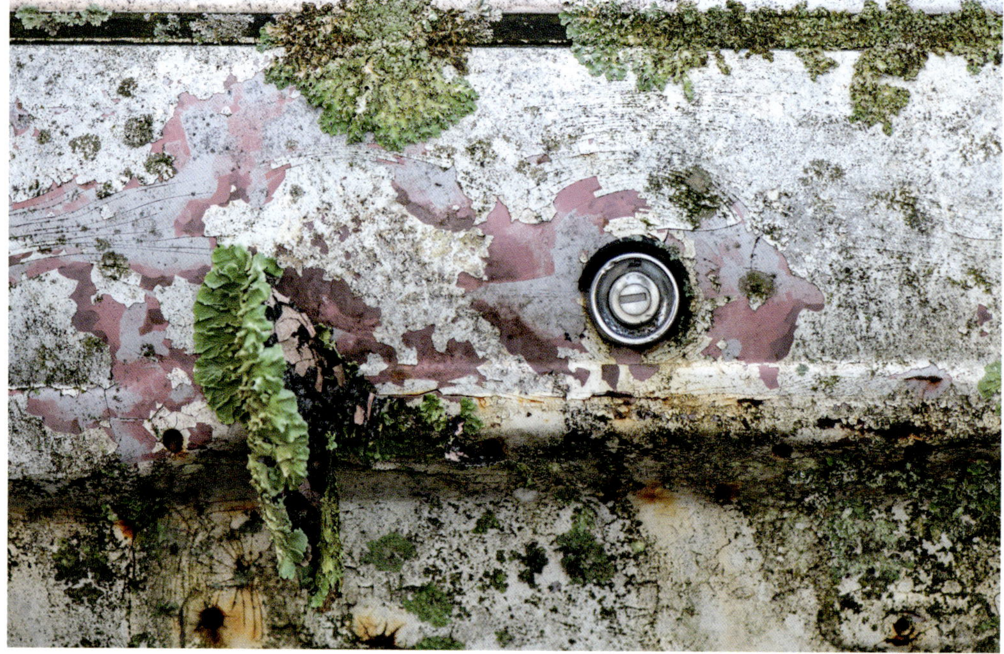

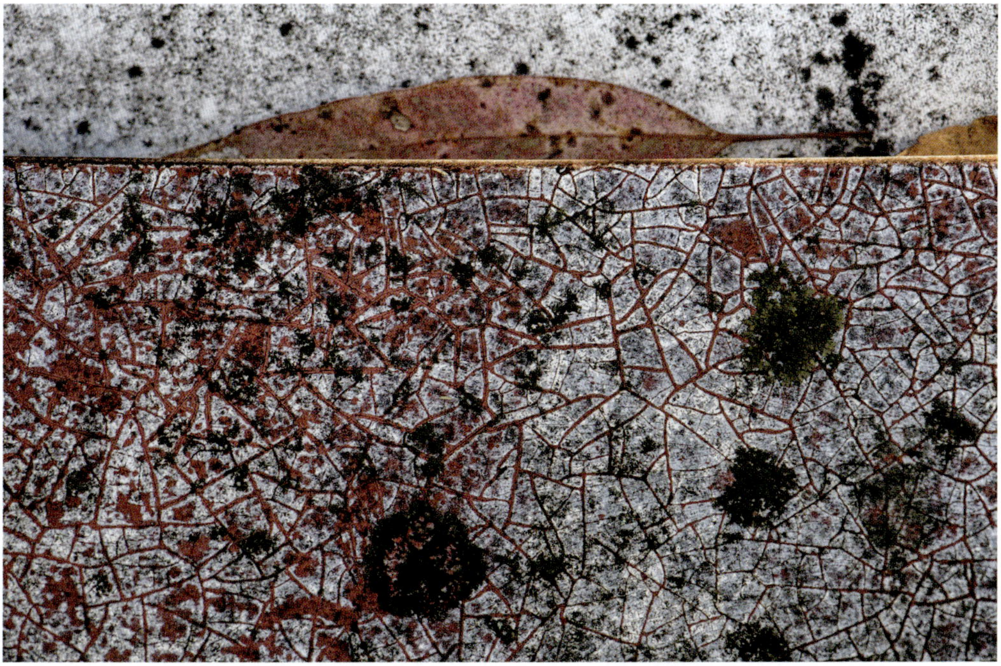

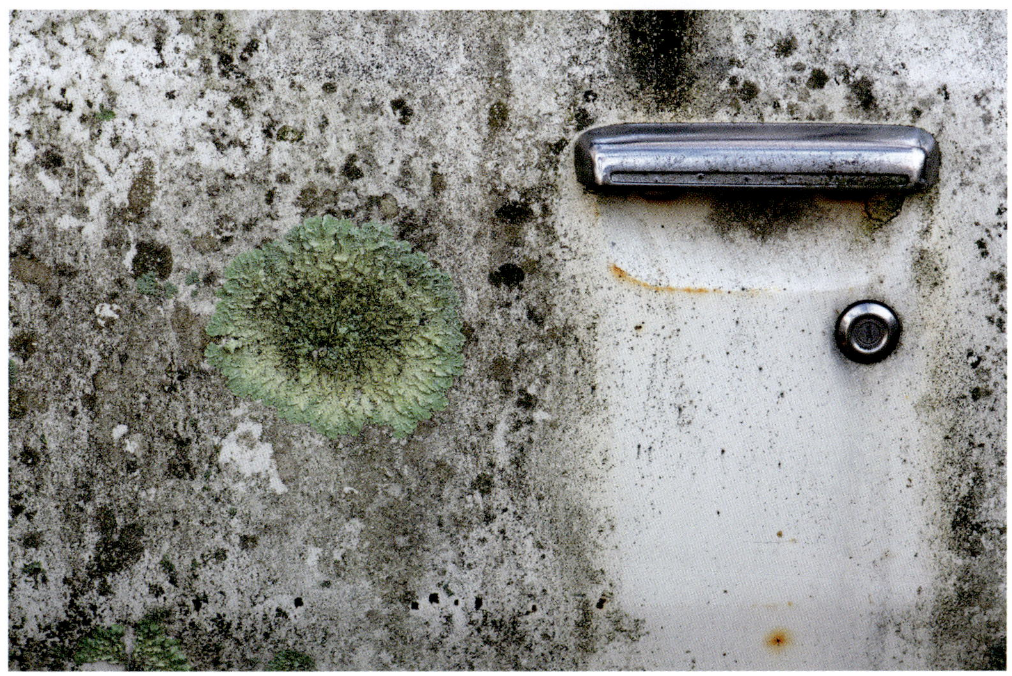

velo-fungus co-habitation

Dryad's saddle
(*Polyporus squamosus*)
and bicycle, Switzerland

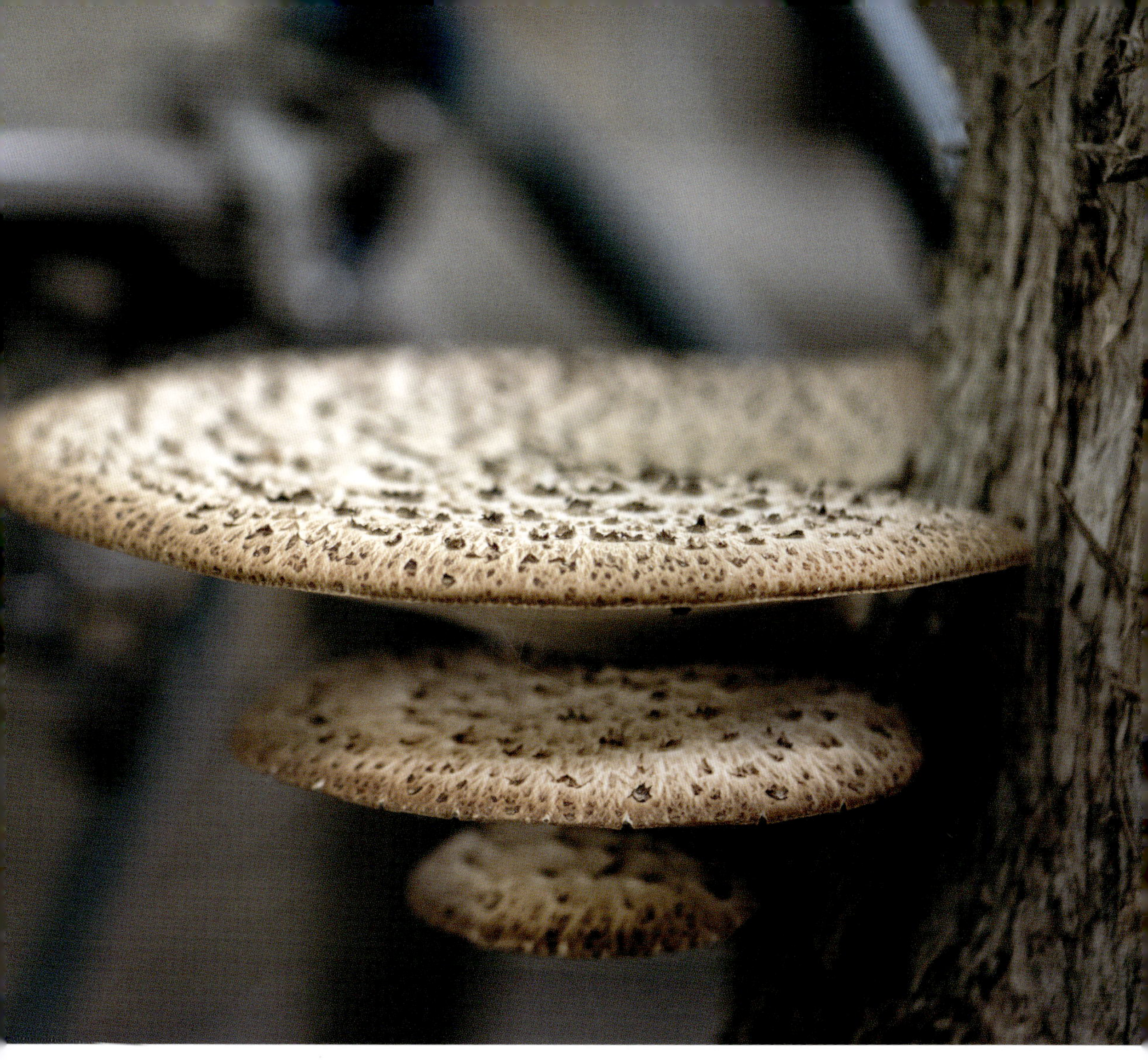

Lichen-encrusted
boiler, Australia

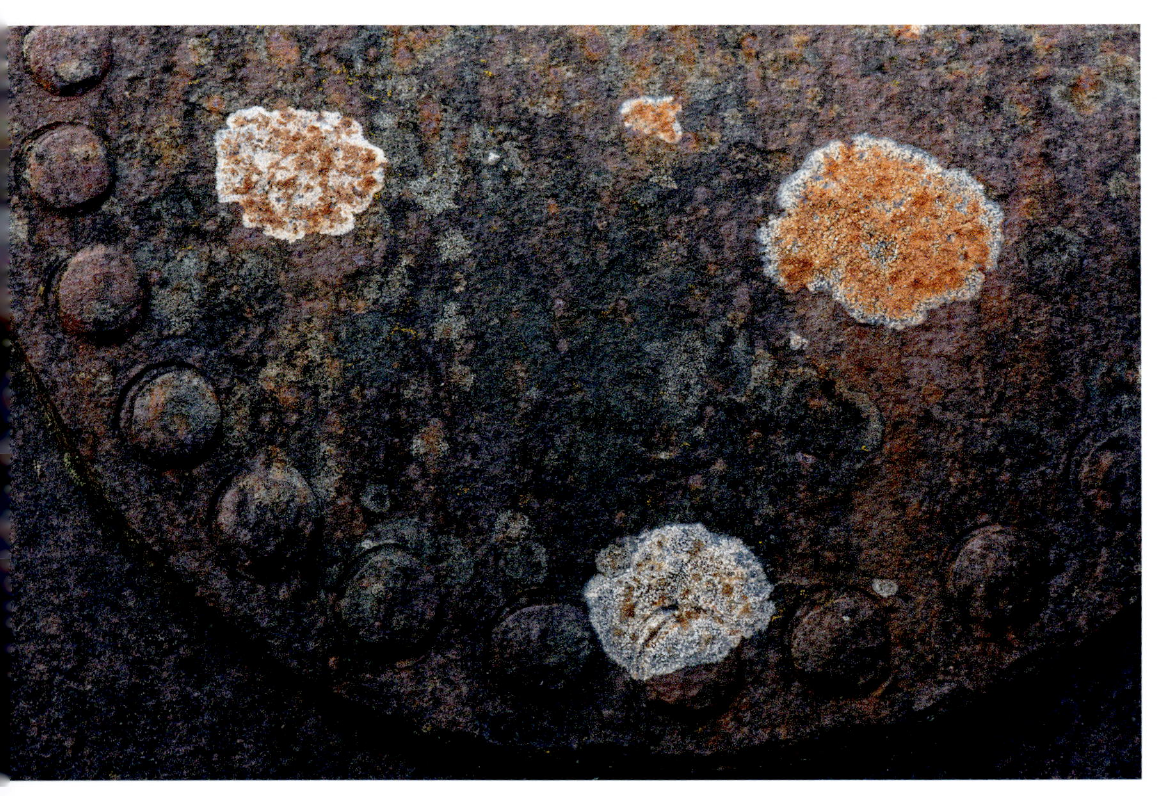

Lichens and moss colonise paving stones, Switzerland

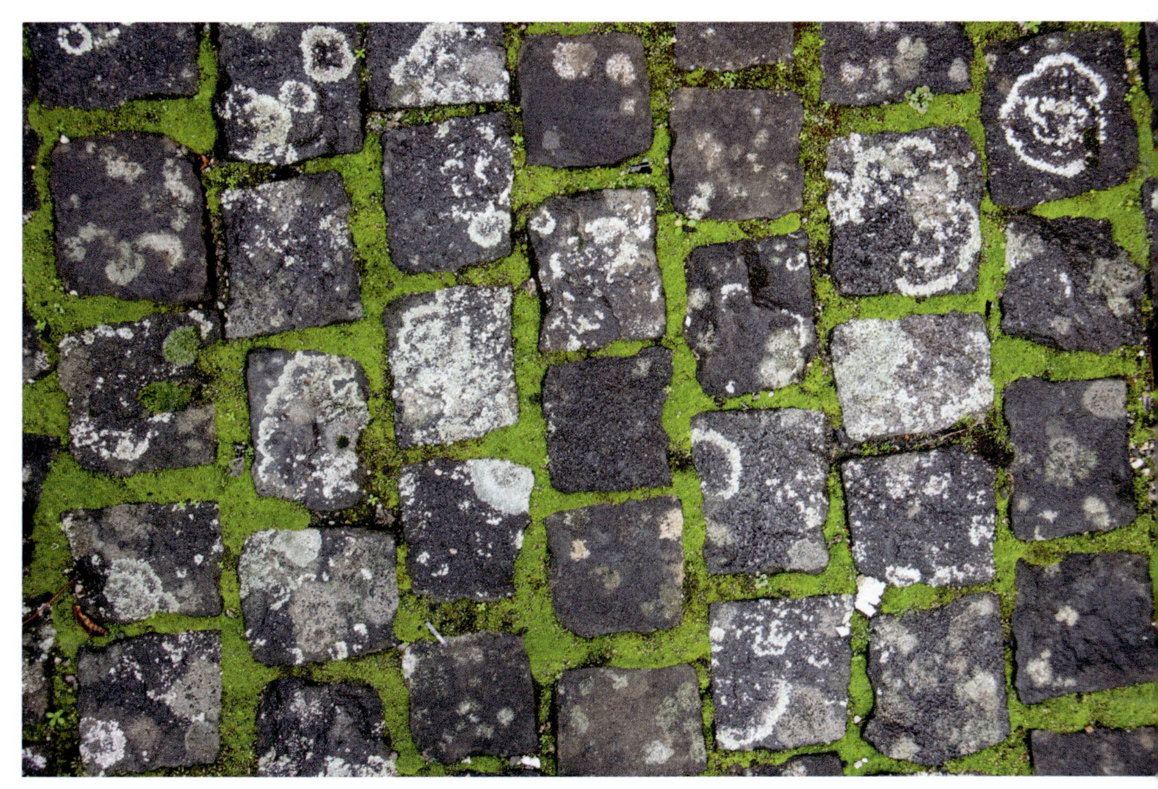

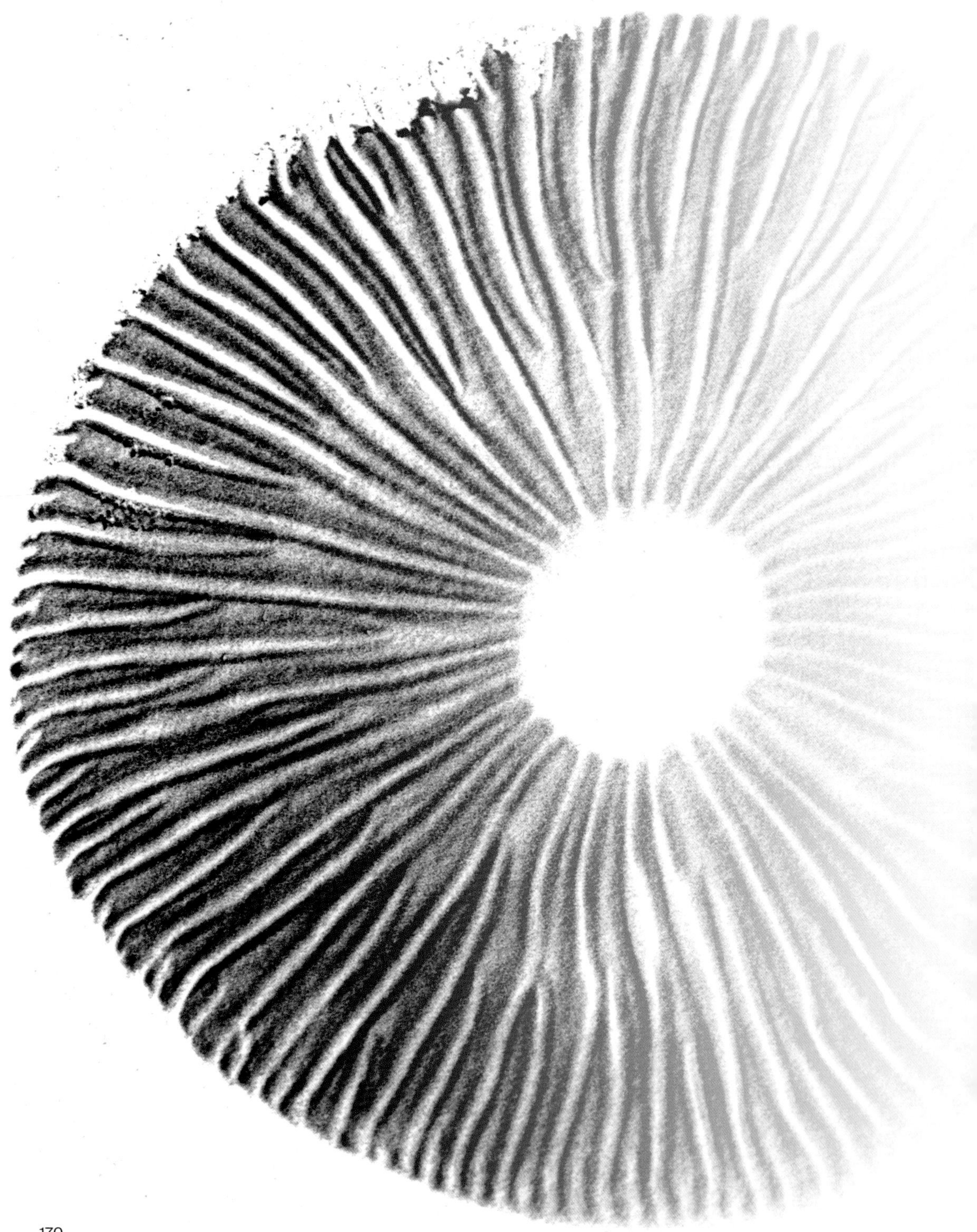

TRANSIENCE

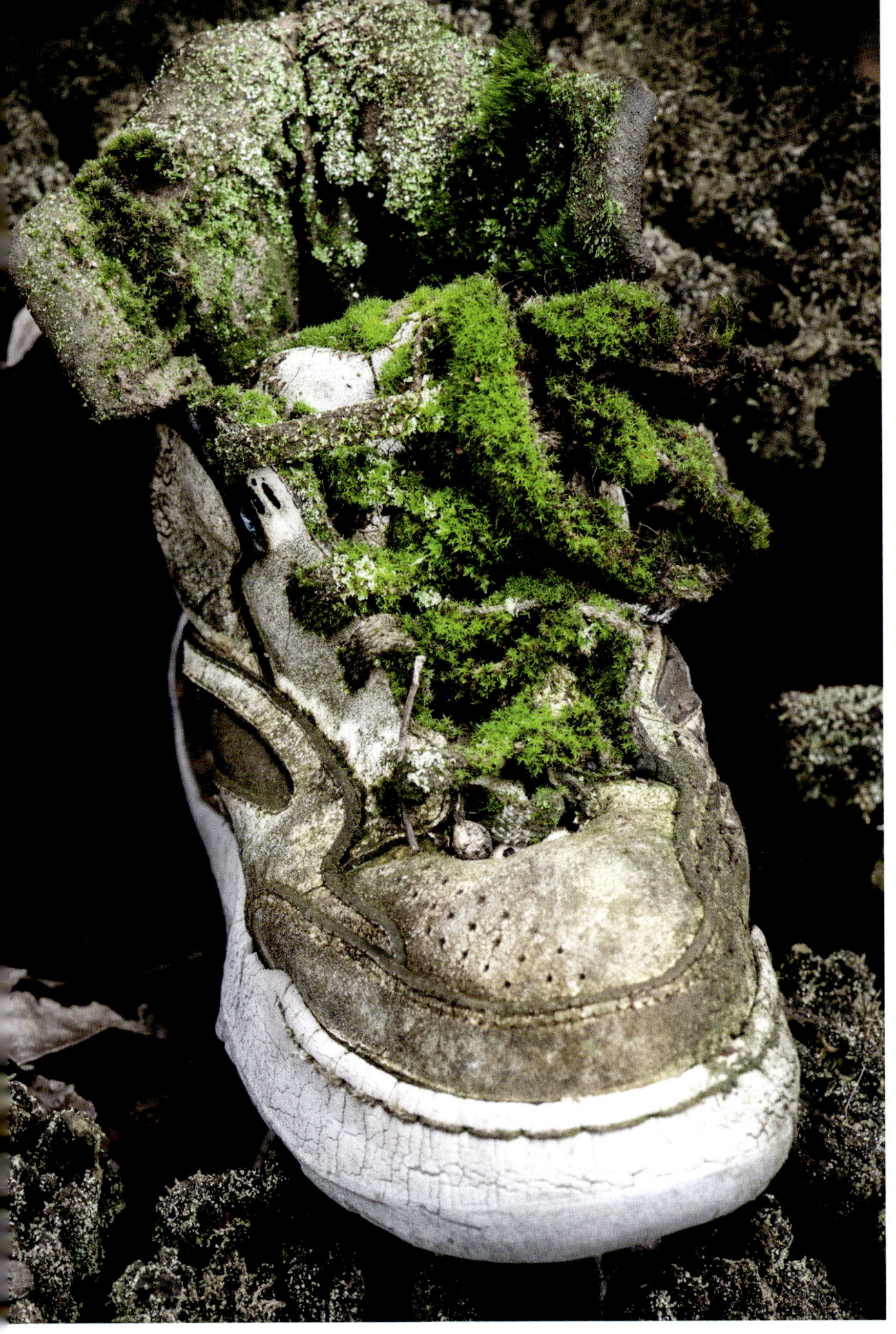

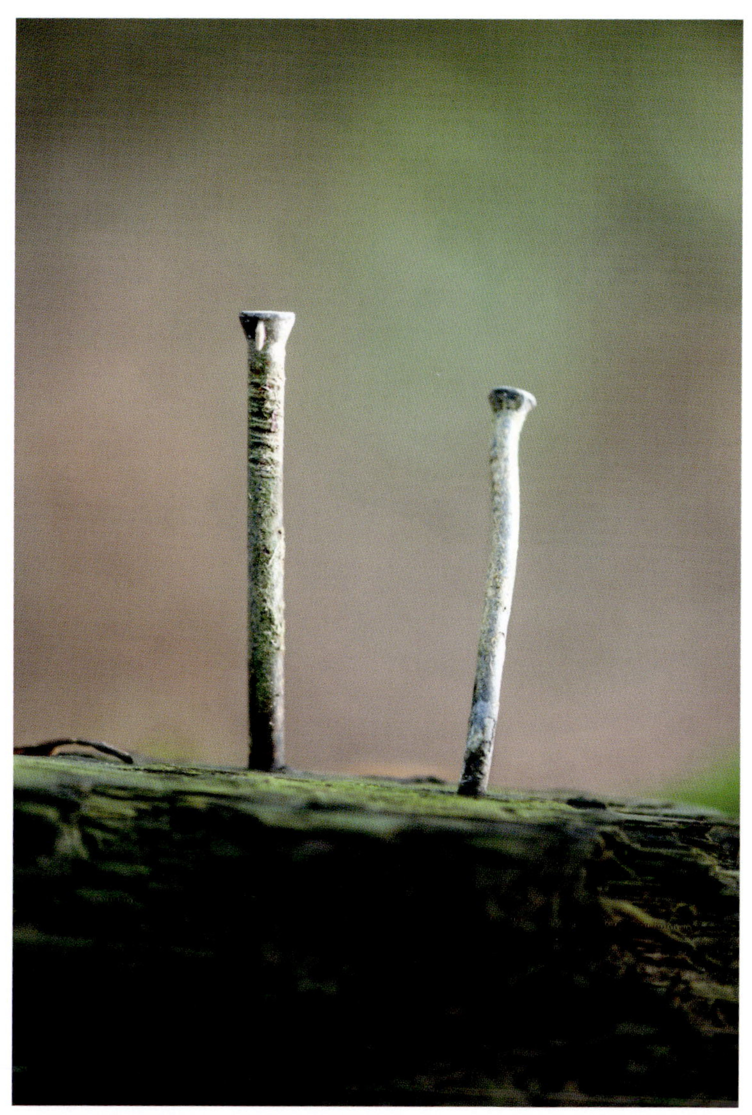

Upheaval

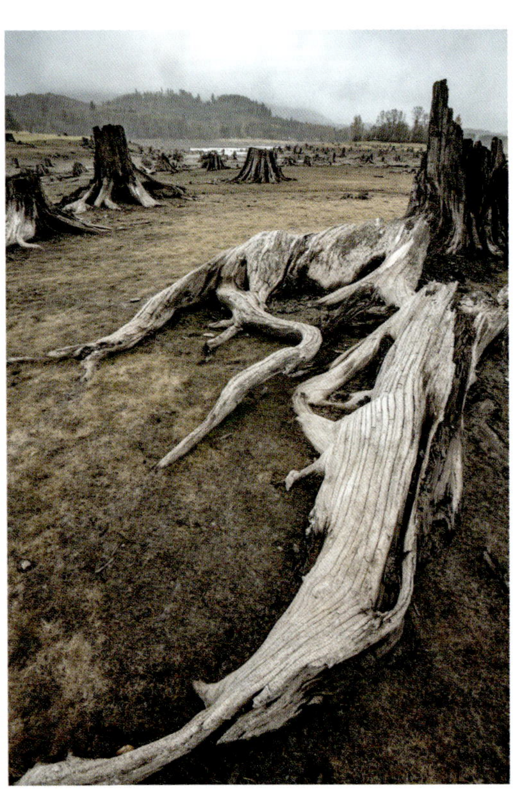

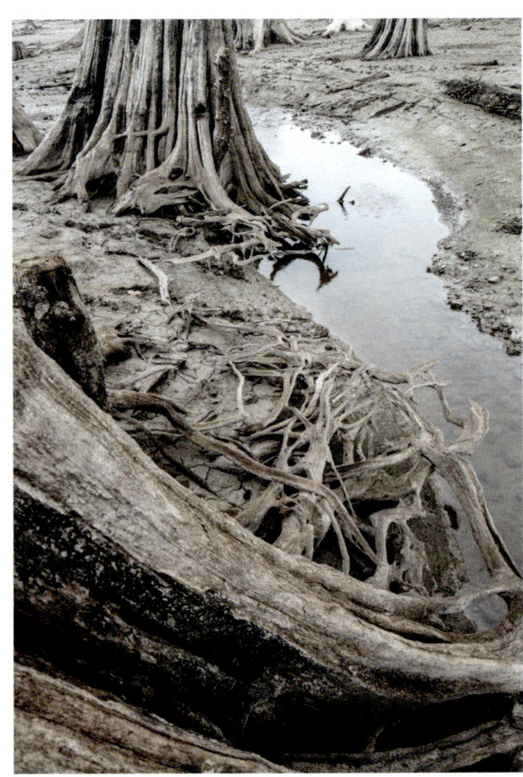

Alder Lake, Washington, USA

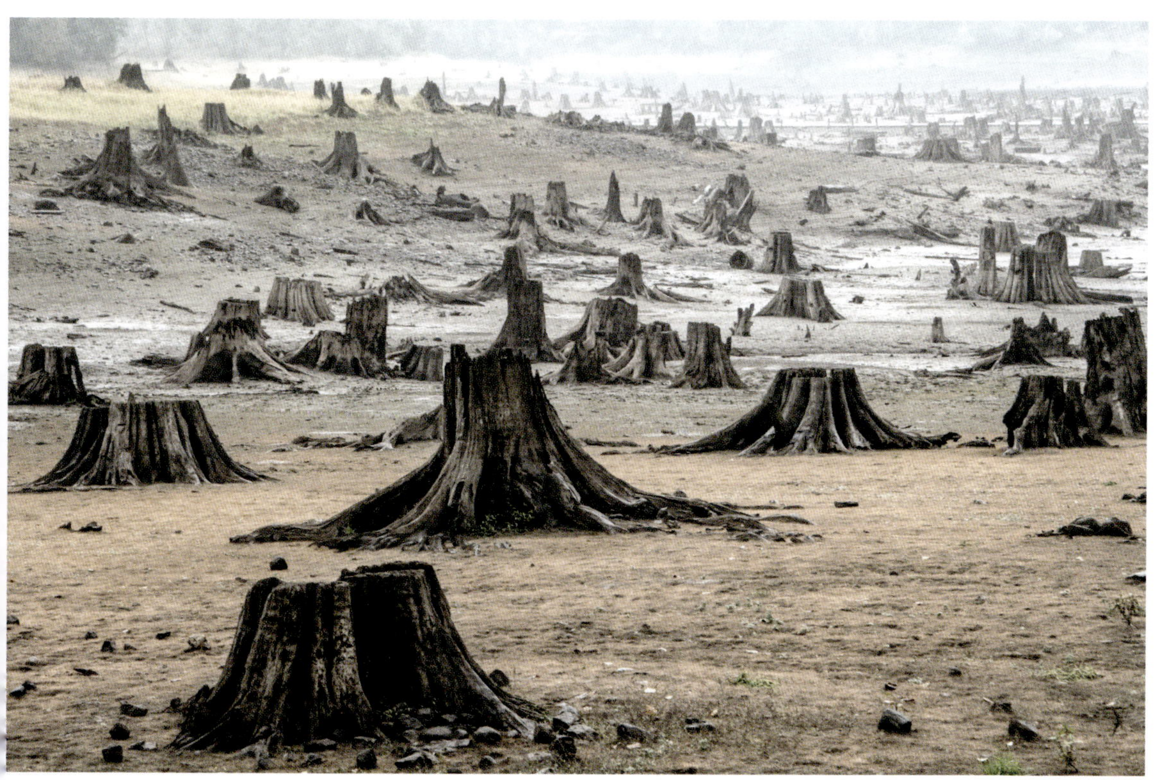

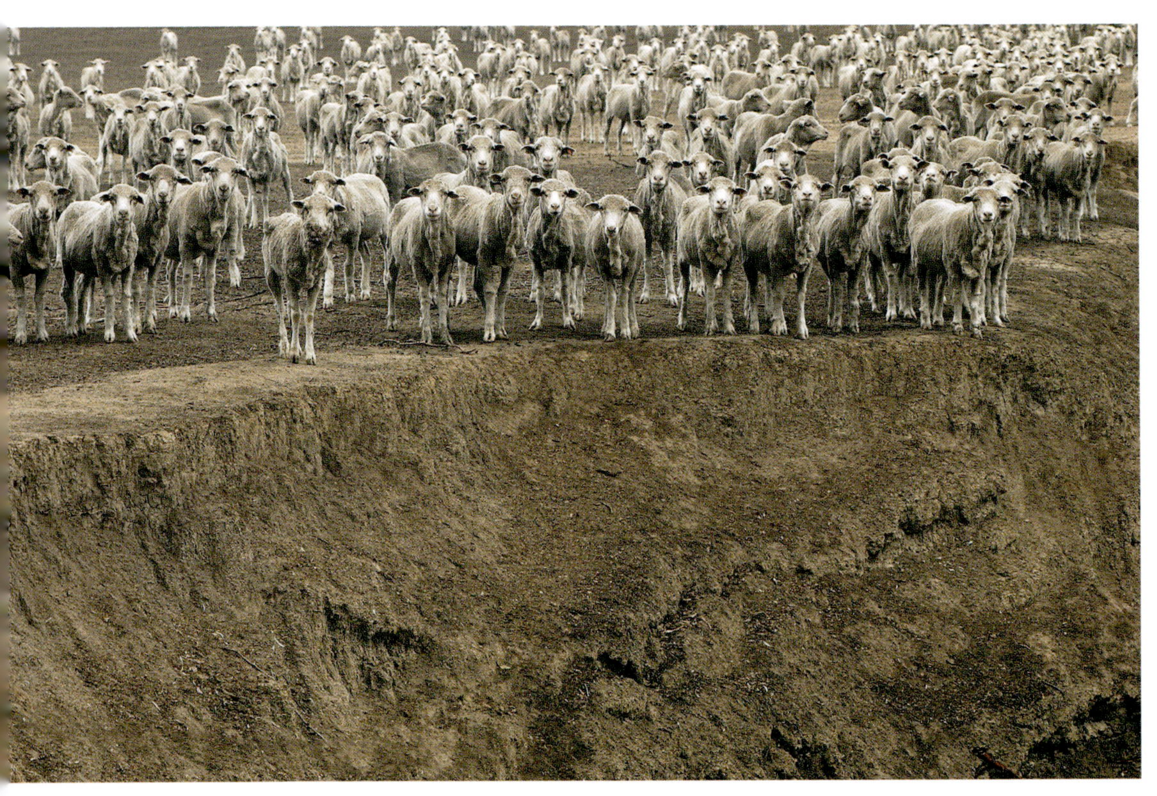

Red Knob, Victoria,
Australia

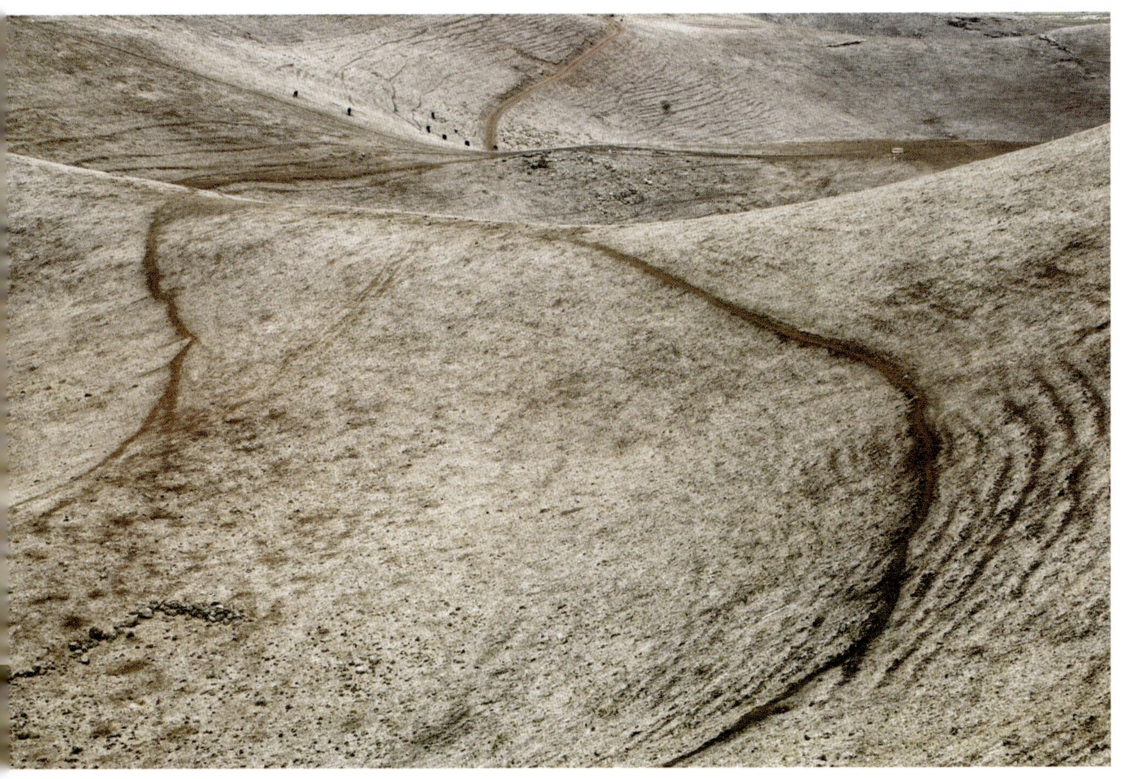

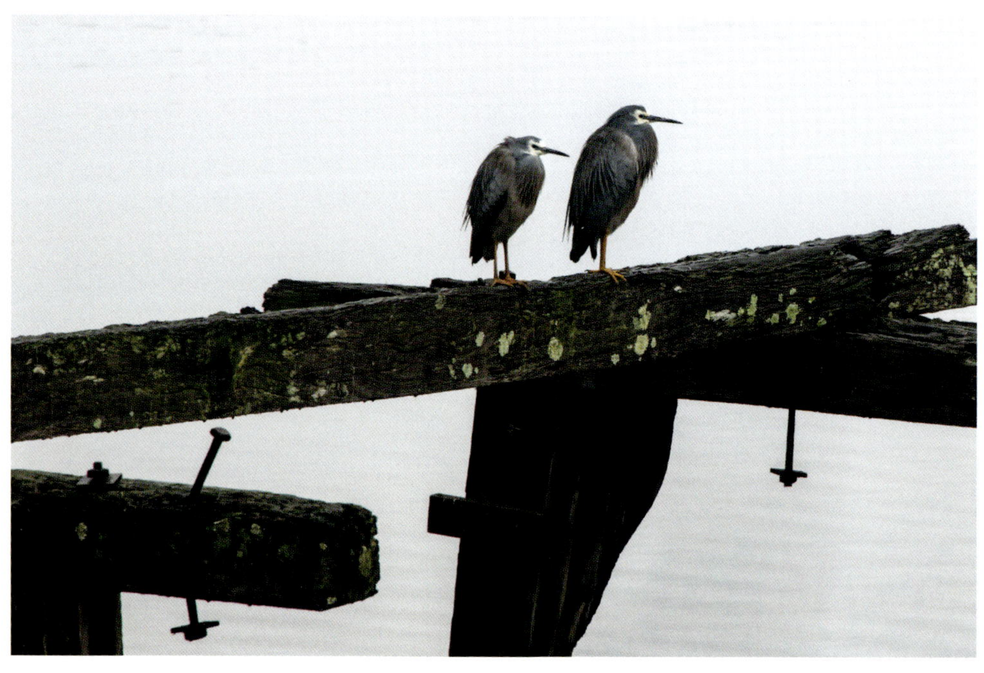

White-faced herons (*Egretta novaehollandiae*), Australia

Epilogue

My viewfinder fogs as I bring my eye in close. Winter isn't far away. The days grow shorter, and the air takes on a distinctive feel and aroma as autumn draws to a close. Squirrels stow hoards of hazelnuts and acorns. Woodpeckers excavate tree trunks and stash nuts and seeds. Throughout the forest, there are signs of stockpiling by other unseen creatures, provisioning for the cold months ahead. Dormice hide out in hollows. Birds are on the move. Bramblings migrate in the millions from northern Europe to overwinter in central and southern Europe and feast on beech-mast. Rainclouds skulk along the horizon as the leaves of deciduous trees slowly swirl to the ground. Beneath the forest floor, mycelium hunker down for the winter. The last mushrooms wither and collapse. Just a few will persist throughout the year. But they'll be back.

I pack my bag for the journey home to Australia, where fungi wait out the hot summer below ground. Many are adapted to extremes of heat and desiccation – and even fire – while others could succumb to altered conditions brought about by a shifting climate. I'll continue to photograph the enigmas of their lives as we all adapt to a changing world, and reflect on my luck in being able to spend my days in the forests among fungi and friends.

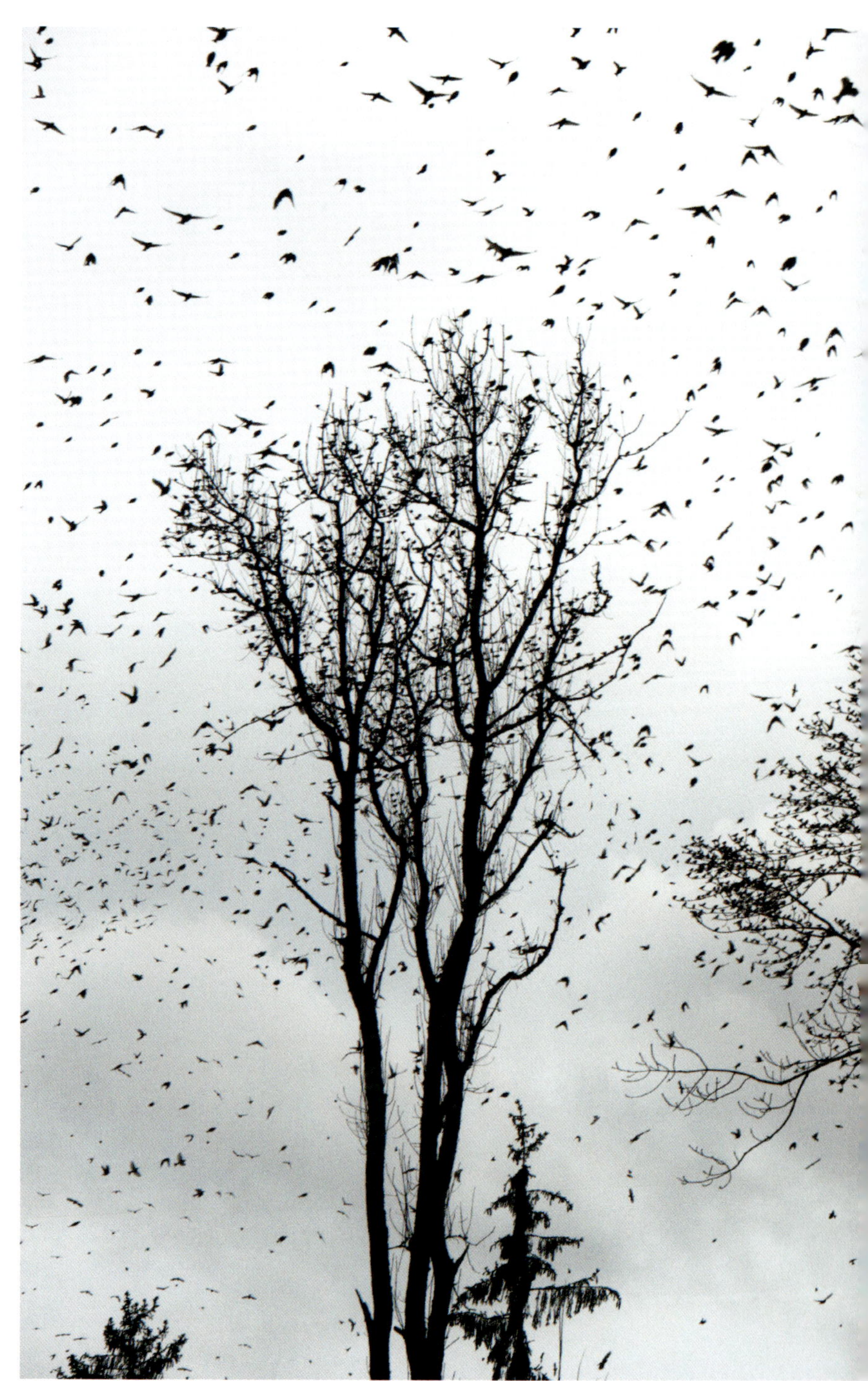

Migrating bramblings (*Fringilla montifringilla*), Switzerland

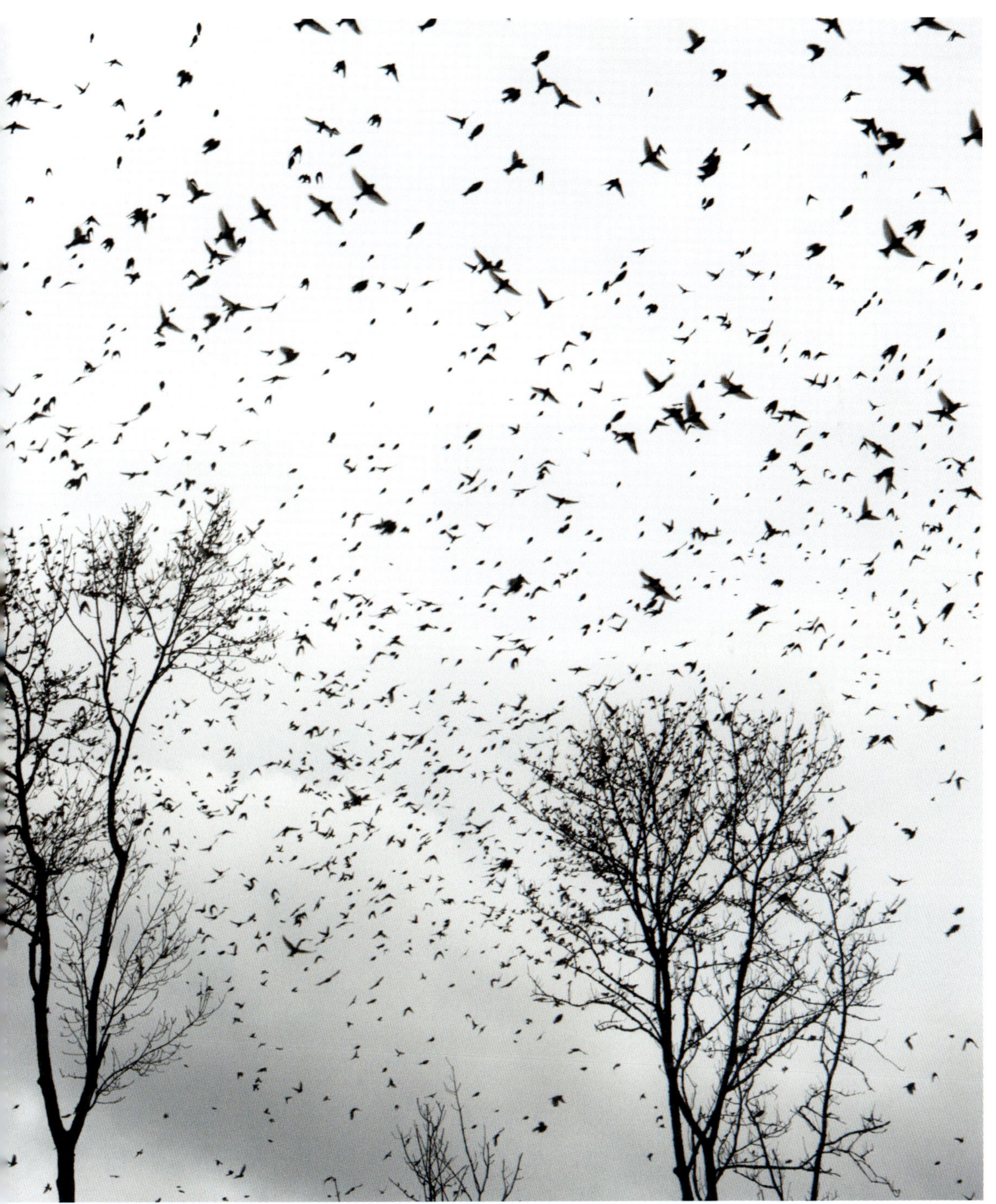

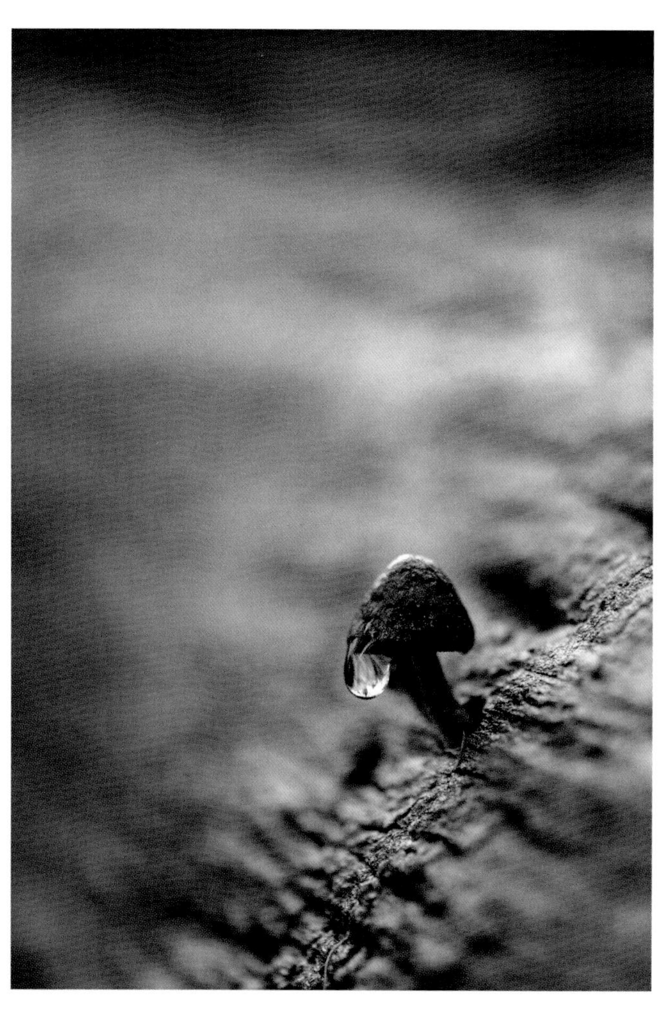

Acknowledgments

Funga Obscura is the result of many years roaming forests in search of fungi. I've shared these journeys with friends and colleagues who offered their insights and ideas along the way. I'm especially grateful to Valérie Chételat for her invaluable assistance with image selection and sequencing, and for travelling on another book journey with me. Heartfelt thanks to Inga Simpson, Kathrin Schlup, Kim Percy, Libby Robin, Marita Smith, Philip Rogosky and Prue Hutton for reading drafts, and Martin Grünfeld for inspiring conversations on the topic of fungal decomposition. Special thanks to Britt Bunyard and Steve Trudell for their generous and astute mycological advice; Sara Calhim, Panu Halme and the Nordic team for their work on the morphological diversity of spores; Franz-Sebastian Krah and colleagues for their research on fungal thermal melanism; and Amira Farzana Samat, Ali Abbas and Dee Carter for their investigation of fungi and plastics. I extend my thanks to all the team at NewSouth Publishing, editor Emma Driver for her encouragement and judicious editing, and designers Klarissa Pfisterer and Hamish Freeman. And every foray – whether in the forest or the lake – is always more fun with Barbara Thüler.

Further reading

Lücking, R & Spribille, T, *The Lives of Lichens: A Natural History*, Princeton University Press, New Jersey, 2024.

Moore, D, *Fungal Biology in the Origin and Emergence of Life*, Cambridge University Press, Cambridge, 2013.

Pauli, L, *Manufactured Landscapes: The Photographs of Edward Burtynsky*, Yale University Press, London, 2003.

Petersen, JH, *The Kingdom of Fungi*, Princeton University Press, New Jersey, 2013.

Pouliot, A, *Underground Lovers: Encounters with fungi*, NewSouth Publishing, Sydney, 2023.

Rubin, R, *The Creative Act: A Way of Being*, Penguin Press, London, 2023.

The University of Chicago Press, Chicago 60637
The University of Chicago Press, Ltd., London

Published 2025
Printed in China

34 33 32 31 30 29 28 27 26 25   1 2 3 4 5

ISBN-13: 978-0-226-84372-8 (cloth)
ISBN-13: 978-0-226-84373-5 (ebook)

DOI: https://doi.org/10.7208/chicago/9780226843735.001.0001

First published in Australia by NewSouth, an imprint of UNSW Press Ltd.

Library of Congress Control Number: 2024948354